COLLECTING HANDKERCHIEFS

ROSEANNA MIHALICK

PHOTOGRAPHY BY CHRISTOPHER CSAJKO

4880 Lower Valley Road, Atglen, PA 19310 USA

DEDICATION

This book is dedicated to my mother and to the spirit of collecting that she has instilled in her children and grandchildren. Thank you for the joy, passion, knowledge, and thrill of the hunt that you have shared with us all.

Published by Schiffer Publishing Ltd.
4880 Lower Valley Road
Atglen, PA 19310
Phone: (610) 593-1777; Fax: (610) 593-2002
E-mail: Schifferbk@aol.com
Please visit our web site catalog at
www.schifferbooks.com
We are always looking for people to write books on new and related subjects. If you have an idea for a book, please contact us at the above address.

This book may be purchased from the publisher.
Include $3.95 for shipping.
Please try your bookstore first.
You may write for a free catalog.

In Europe, Schiffer books are distributed by:
Bushwood Books
6 Marksbury Ave.
Kew Gardens
Surrey TW9 4JF England
Phone: 44 (0)208 392-8585;
Fax: 44 (0)208 392-9876
E-mail: Bushwd@aol.com
Free postage in the UK. Europe: air mail at cost.
Try your bookstore first.

Copyright © 2000 by Roseanna Mihalick
Photography by Christopher Csajko
Library of Congress Catalog Card Number: 00-100202

Book Design by Anne Davidsen
Type set in Lithograph/ Souvenir Light

ISBN: 0-7643-1131-X

Printed in China
1 2 3 4

CONTENTS

ACKNOWLEDGMENTS

In the words of the *Beatles*—"*I get by with a little help from my friends...,*" this book could never have been completed without the help and support of my friends and family. Their good wishes and encouragement have meant so much to me. My family, and especially my husband, Peter, have been terrific. When a busy "Mom," collector, dealer, and group-shop member sets out to write a book, something has got to give! In this case everything and everyone did. They gave me time and patience, tolerance and understanding. They really proved themselves.

My mother's collection is widely represented here and she has been a source of suggestions and information. In fact, it was she who came up with the idea for the book when we met editor Doug Congdon-Martin at a church supper during the Brimfield shows. I would like to thank my editor and Schiffer Publishing for the opportunity to do this project.

The wonderful photos in this book are the work of a very talented young man, Chris Csajko. Chris is an ambitious photography student at the School of Visual Arts in Manhattan. He handled this assignment in a totally calm and professional manner.

I am not a typist so the task of translating the manuscript into a finished product fell into the very capable hands of Eileen Shields. How she managed while caring for her four young children and working, I'll never know! I am both impressed and grateful.

Fellow dealer/collector Mary B. Ross was a ray of sunshine. She filled in the gaps in this book with some terrific examples from her own collection.

So many others have offered a kind word and encouragement that it would be difficult to mention them all. Please know I do appreciate all of you!

INTRODUCTION

When did the handkerchief, originally meant for utilitarian purposes, become an item worth collecting? And why are some handkerchiefs more worthy of the label "collectible" than others? These are very important questions, and they are essential to the purpose of this book.

Many collectors would agree that handkerchiefs became worth collecting simply because they aren't widely produced any more. The more discerning will assert that not every hankie is worth collecting. However, many of us save a special handkerchief for a sentimental reason—a gift from a dear aunt, a souvenir of a memorable occasion, or a memento from a loved one. According to that criteria, any hankie is worth preserving.

Yet, the bright colors and intriguing designs of the handkerchiefs of the 1940s, 1950s, and early 1960s cause many collectors to consider these the most desirable. The true collector will search for fine quality in the fabric of the handkerchief, a hand-rolled or neatly machine scalloped edge, a fine lace trim, creativity of design, and original paper labels or tags.

The history of the handkerchief and its social significance has been widely studied and I make no attempt to take on that scholarly task. Also, J. J. Murphy's fabulous book *Children's Handkerchiefs* published by Schiffer Publishing Ltd. has thoroughly covered that topic. This book will not deal with those rare and priceless eighteenth and nineteenth century lace-edged handkerchiefs found in museums around the world, but rather with those mass-produced women's handkerchiefs of more modern times. The era following WWII and into the 1960s offers so many wonderful, fun, and original designs that you might call it the "Golden Age of Handkerchiefs." Like so many things that are brightest just before they fade out, this brief time period produced some handkerchiefs that are definitely worthy of collection. The "Hankie era" ended with the arrival of the more sanitary paper disposable tissue, in the late 1950s.

My mother's extensive collection has served as the basis for this book. Although I have been collecting for several years, the variety of her collection far surpasses mine. Since we often shop together, a friendly rivalry exists between us to find that "special" treasure. We find our hankies all over the country at garage sales, flea markets, and antique shows or shops. There is in fact one small show in New Jersey that we jokingly refer to as "hankie heaven" because it never fails to yield a new item for our collection. We want to dive right in when we find a likely stack of handkerchiefs, but a methodical approach works best and dealers really appreciate it when merchandise is returned to its proper order.

Also, a word of advice to dealers—sticky price tags leave glue marks and hankies folded and stored in plastic bags are unattractive. The folds cause fabric stress and the plastic bags trap moisture, both of which will eventually destroy the item. The best way to show a handkerchief is in a full open position, lying flat, free of fold marks, and crisply pressed. A small price tag pinned in a corner is preferred.

If you are storing your handkerchiefs, use acid-free tissue, boxes, and folders. Framing is a great way to display your handkerchiefs, but use acid-free mats and UV protective glass that doesn't make contact with the fabric. Laundering can cause problems, so if you must, use care—hand wash and dry flat.

No book about collectibles can be considered complete without a price guide, but remember it's a guide and prices are subject to many variables. Since I live in the New York metropolitan area, the prices offered reflect that. Also, this is a fairly new area of interest and prices are necessarily speculative at times. Condition is noted in the captions with wear or stains decreasing the value. The author and publisher accept no responsibility for the losses or gains related to the prices quoted in this book.

CHAPTER 1
WHITES & WEDDINGS

The ubiquitous white handkerchief—clutched in *every* bride's hand or discreetly tucked in *every* lady's purse—is represented here by many unique and several fairly common examples. Although stark in color, the fine detail and warmth of lace make the *white* handkerchief a truly beautiful part of *every* handkerchief collection. It is natural, therefore, to begin this book with the seemingly simplest of all hankies, plain white.

It becomes obvious, after some observation, that white handkerchiefs are anything but plain. In fact, some of the most elaborate designs and most elegant lace and trims appear on simple white linen and cotton squares. Just as no woman of the past would consider her wardrobe complete without several white handkerchiefs, no collection of handkerchiefs can be complete without several snowy white examples of handkerchiefs in their highest form, the wedding handkerchief.

The bridal handkerchief truly raises the handkerchief to its epitome. The laces used are the most valued and the fabric is the finest in quality. They are the family treasures that will be handed down from mother to daughter. They are the heirlooms that will survive for generations to be used on that special day, often becoming the "something old" of the bridal litany—"*Something old, something new, something borrowed, something blue…*"

Also included in this chapter are the lovely white-on-white handkerchiefs that were once the needful things of a woman's daily life before the disposable tissue came to be.

The punchwork technique is used to embellish this white wedding handkerchief in an all-over floral design. Only a small center remains unadorned, c.1920. 10" linen. $30-45.

Another punchwork example, the design almost completely covers the square in a dramatic geometric pattern. Small-embroidered flowers complete the motif, c.1920. 10" linen. $30-45.

This tiny handkerchief has an intricate border of drawn-work with Maltese crosses in each corner. The edge is very fine old Valenciennes lace. Hem stitch. This is a "one-of-a-kind," all handmade, c.1900. 8.5" linen. $50-60.

Machine-made lace edges this example. The flowing curves of the lace are echoed in the shape, which makes it unique. Original label—Made in Austria, c.1950. 11.75" linen. $30-45.

The 2-inch deep lace border on this bridal handkerchief is probably Valenciennes. The corner medallions are a type of chemical lace. Softly used, c.1920. 11" linen. $35-50.

The original label tells us that this lovely piece was made in Austria. The 2-inch floral lace border is attractive and feminine, c.1950. 11.5" linen. $30-45.

A rare example, fine handmade lace medallions fill the corners of this handkerchief. Entirely hand done, the lace is of the "free bobbin" type, with small flowers connected with braids or "brides," as they are sometimes called. The edge is of the same type attached with exquisitely fine stitches, c.1900. 12" linen. $60-75.

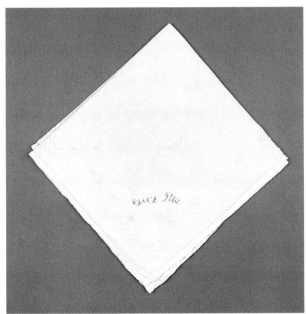

White-on-white dotted Swiss type embroidery. This is in the collection because of the enigmatic message/name in indelible ink, c.1930. 10" linen. $10-15.

A lovely machine-made piece in an ever-popular white theme. Large florals in one corner with smaller motifs in the other three. Suitable for special occasions, c.1930. 10" linen. $10-15.

Heart shaped leaves appliqued on an embroidered vine with a fine hemstitch border. The design appears in each of the four corners, c.1940. 11" linen. $15-20.

A crocheted edge and "pulled-thread" decorate this simple piece. The obvious hand done quality, with uneven stitches, adds to the charm of this handkerchief, c.1930. 11.5" linen/cotton trim. $20-30.

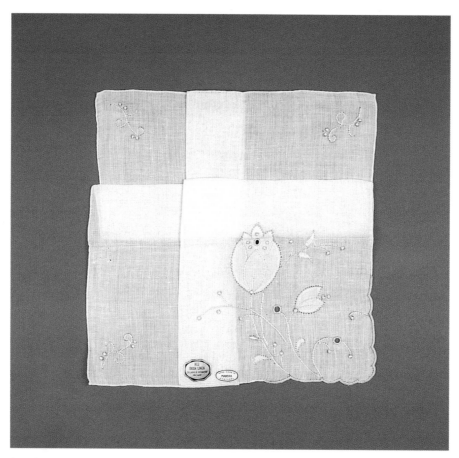

A stylized tulip appears in one corner and small tendrils are in the other three.
The original folds and labels remain, c.1950. 12" linen. $10-15.

A scalloped edge and floral Swiss style machine embroidery are finished by a geometric hemstitch design on this white work hankie, c.1920. 11.5" linen. $15-20.

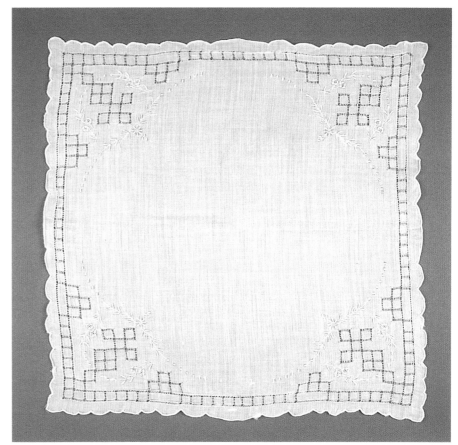

The simple elegance of the geometric border with repeating bars of embroidered flowers make this a special handkerchief, c.1930. 10" linen. $15-20.

The charming blue tinted work associated with Madeira embroidery is evident in this square with symmetrical patterns all around its surface. Evidence of the paper label remains, c.1930. 9.5" linen. $15-18.

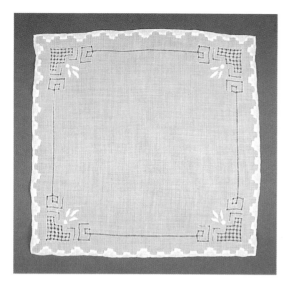

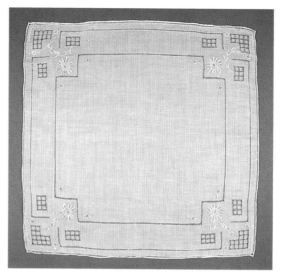

Handmade "puled-thread" corners and a detailed
hem make this a worthwhile item for a collection,
c.1930. 11" linen. $18-25.

Open-work designs frame the Swiss style embroidery
of this simple white handkerchief, c.1930. 11" linen.
$15-20.

The 3-inch deep border of machine-made lace on this tiny square
makes it well suited for the bride. Evidence of original paper label
remains, c.1930. 11" linen. $25-30.

Above: With medallions of fine "chemical lace," this piece has the look of fine bobbin lace at a fraction of the cost, c.1930. 11" linen. $20-35.

Below: Shadow work of white-on-white daffodils and a deeply ruffled edge are dramatic. The beautiful floral silhouettes on this oversize handkerchief make it suitable for a collection, slight wear, c.1950. 16" cotton batiste. $20-30.

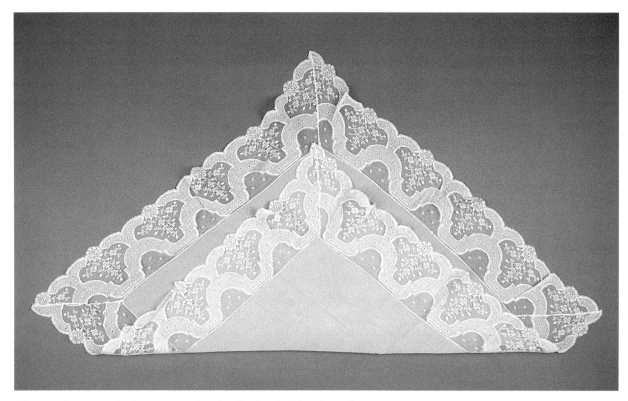

This is a dramatic off-white evening handkerchief with 2.5 inches of fine lace trim. A must-have for the lady of fashion and any collector. Some wear and stains, c.1940. 16" silk chiffon and lace. $30-40.

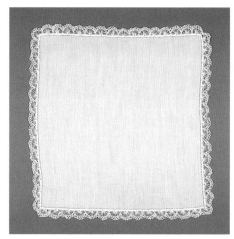

The knotted lace edging, entirely hand done, makes this simple piece worth owning. This type of work is called "Armenian Lace" due to its Middle Eastern origins, c.1950. 10" linen $30-40.

A deep lace border is formed by bands of lace decorated with grape clusters. Soft and delicate, some wear, c.1910. 12" linen. $25-35.

The edging is a fine example of knotted lace worked with a simple sewing needle. This airy trim is similar to pin lace, c.1950. 10" linen. $30-40.

The tatting used to trim this handkerchief is very elaborate and done by a fine hand. The insert in one corner is lovely, c.1940. 12" linen. $35-45.

The tatting that trims this square is simple, but beautifully done. The original box folds remain, showing it was never used, c.1950. 12.5" linen. $25-30.

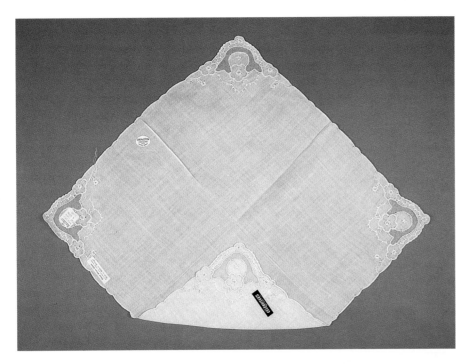

This handkerchief is unusual because it retains all its original labels—Made in Portugal, and an original price of $1.25, c.1950. 12" linen with cotton embroidery. $25-30.

The shamrock-covered lace trim made this a must for collecting, c.1950. 10" linen. $15-20

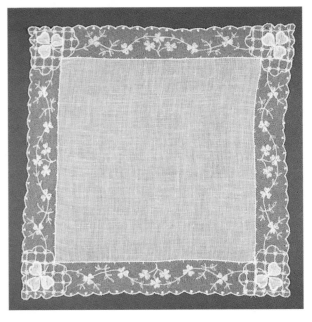

Enriched by the trim of true handmade bobbin lace, this white handkerchief also has a fine pulled thread hem. The corner design is a geometric grid and dots, c.1920. 9.5" linen. $30-40

The punchwork or "buratto" technique is used here to produce a beautiful floral pattern. Often used on bed and table linens, the technique produces an artful design on this handkerchief, c.1930. 11" linen. $25-35

Last, but certainly not the least, this beautiful handkerchief is a show-piece from the Brussels Exposition. A fantastic example of fine Brussels lace, c.1958. 9" linen/cotton. $50-60

CHAPTER 2
CALENDARS

A collection of calendar handkerchiefs provides a sure way to trace the development of design and style in handkerchiefs. By viewing a collection in chronological order one can observe the rise and fall of the "golden era" of handkerchiefs.

Our collection begins in 1951 and moves forward in time to 1969. The 1951 example exhibits a vibrancy of design and shows a bold use of color. The artwork echoes the style of the 1940s, but the circular design has movement and excitement. This sense of excitement and fun is apparent into the designs of the late 1950s. Once the '60s begin, you sense a shift and notice a decline in quality. The colors become more predictable and the styles are repetitive. The handkerchief isn't a fun accessory anymore, just an old leftover from a former time.

The first calendar handkerchief purchased for the collection, 1951.
(It's the year I was born!) 13" cotton. $20-25

1952 is missing, so we jump forward a year. Brown background with a circular design, 1953. 13" cotton. $20-25

A white background and a vivid red bow as a border enliven this calendar surrounded by a turquoise machine scalloped edge, 1954. 13" cotton. $20-25

This design is based on the popular colors of the time—brown, beige, pink, and turquoise. A more "modern" rendition of the year's calendar, 1954. 13" cotton. $20-25.

This young lady seems to be enjoying her year. The gray background serves as a foil for red, green, and yellow designs with a scallop edge, 1955. 13" cotton. $20-25.

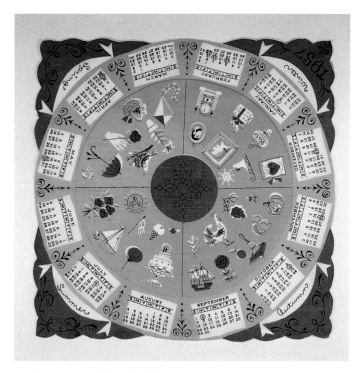

Still hunting for 1956—the following year acknowledges the four seasons in turquoise, powder blue, chartreuse, and orange. Scalloped edge, 1957. 13" cotton. $20-25.

Fashionable young women parade through this year. The color printing is slightly off register. Machine edge, 1958. 13" cotton. $20-25.

Two different color versions of this calendar. Red and blue backgrounds. Machine scallop, 1959. 13" cotton. $20-25.

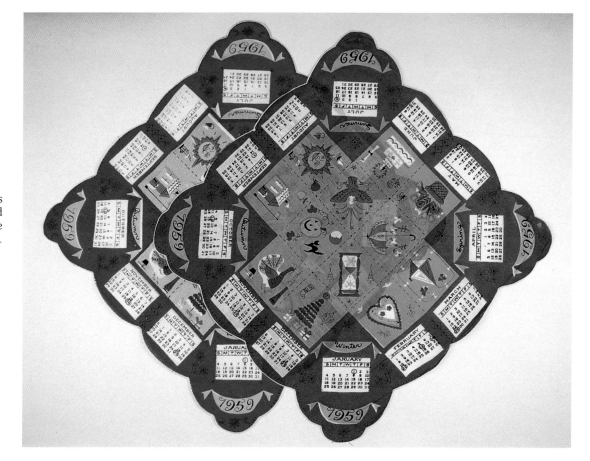

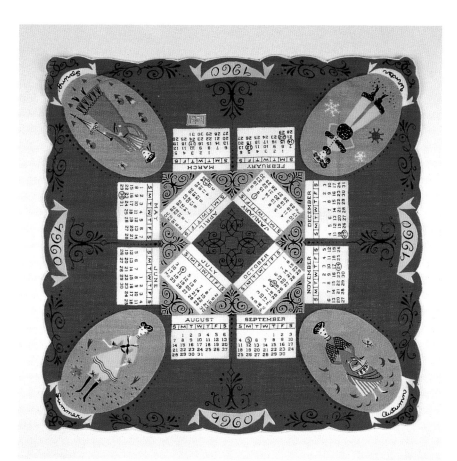

Times were changing rapidly in the 1960s and 1890s themes seemed quaint. The red background highlights a young lady dressed in the fashion of that older time. Original label. Machine scallop edge, 1960. 13" cotton. $20-25.

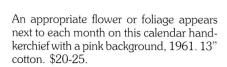

An appropriate flower or foliage appears next to each month on this calendar handkerchief with a pink background, 1961. 13" cotton. $20-25.

A FLORAL THEME DOMINATES THESE MID AND LATE '60S EXAMPLES. ALL ON WHITE BACKGROUNDS. THE LOSS OF CREATIVITY OF DESIGN AND COLOR SHOW IN THESE CALENDAR HANDKERCHIEFS MADE IN THE U.S.A., C.1960. 13" COTTON. $15-20.

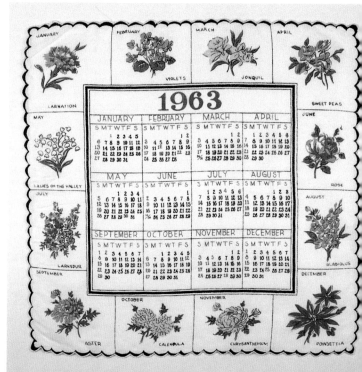

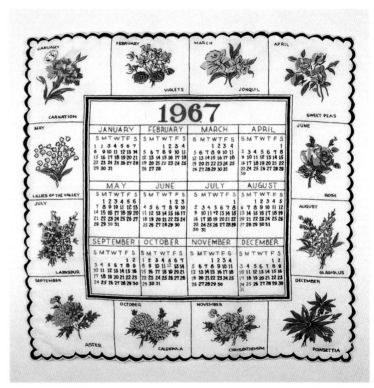

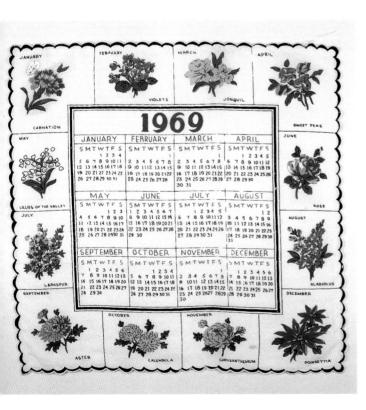

The deep blue background suits the European Christmas theme of these calendars. Suitable for stuffing in a holiday stocking. Original labels—Made in Switzerland, 1965-1966. 12" cotton. $15-20.

Most of the color and fun are gone from these late 1960s examples. The eclipse of the handkerchief era is at hand. Made in Switzerland, 1967, 1968, and 1969. 12" cotton. $10-15.

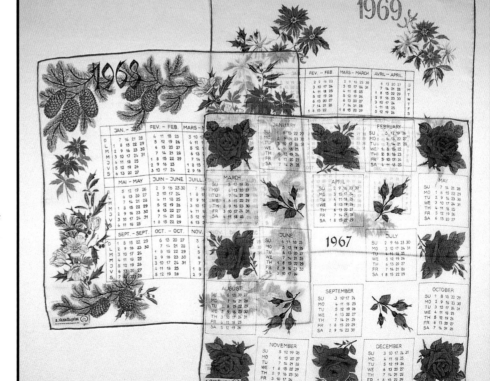

No year is assigned to this "trés charmant" handkerchief done in French by Peynet. This designer-signed calendar chronicles a year of romance, c.1950. 15" linen. $25-30.

CHAPTER 3
HOLIDAYS

It seems as though there is a handkerchief to celebrate every holiday of the year! In fact, these festive squares were very much a part of everyday life. They were not rare or unusual in their own time, but were as common as a greeting card is now. In fact, they often accompanied a card as a small gift or memento.

Depictions of the important days on the calendar have become very collectible in recent times and recall fond memories of seasons past. We have yet to obtain a Halloween or Easter hankie, but are always searching and adding to the collection. Christmas and Valentine themes are easiest to find, so look for examples with unique colors or designs.

A nod to the New Year! Images of a dancing couple, musician, and a clock share the corners with "Baby New Year" in this festive example, c.1953. 15" linen. $20-25.

Love notes carried by doves from him to her and her to him. Vivid red, white, and black make a dramatic modern Valentine, c.1950. 11" linen. $20-25.

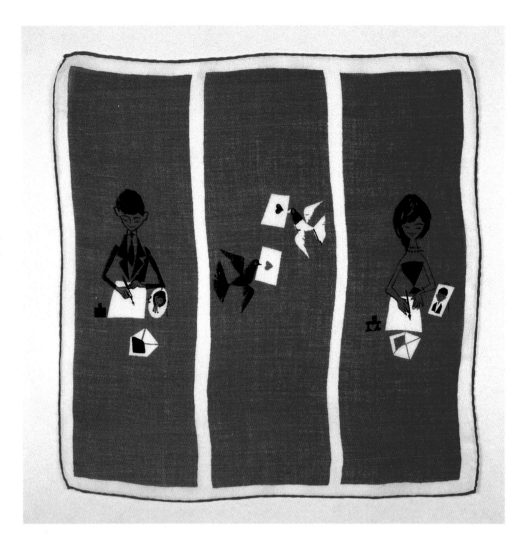

Designed by Pat Prichard the central clock rings in the New Year, but is that cupid in the corners with turquoise hearts? A bit confusing, perhaps a combination of Valentine's Day and New Year's? c.1950. 14.5" linen. $20-25.

This cute cupid is a 1950s version of the winged god. Bright color contrasts make this an interesting Valentine theme hankie, c.1950. 12" linen. $20-25.

Printed on cotton batiste and covered with Valentine motifs, these heart decorated beauties bring delight with their bright colors and ruffled edges, c.1950. 14" and 15" cotton. $15-20.

The crisp contrast of red-on-white is done in cotton machine embroidery, c.1960. 10" cotton batiste. $10-15.

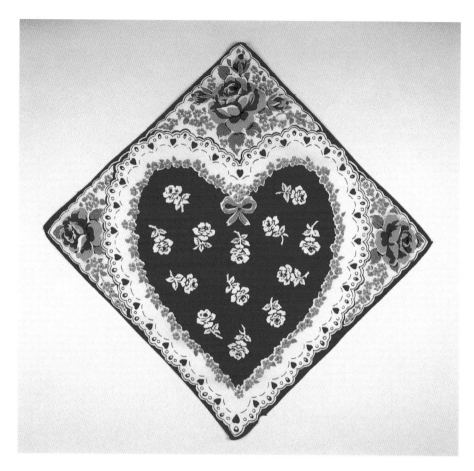

A charming design with roses for remembrance and blue forget-me-nots. Slight edge damage, c.1950. 12" cotton. $10-15.

A rare round Valentine handkerchief. It has the expected hearts, roses, and forget-me-nots in a lovely circular design. Some stains. 15.5" cotton. $20-25.

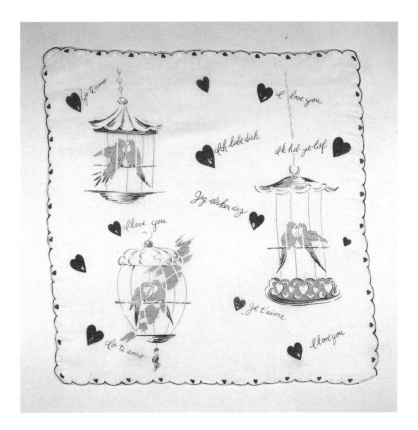

The love birds know that the message "I love you" is the same in any language. A very charming contemporary styled Valentine hankie, c.1950. 14" cotton. $20-25.

Designers of the 1950s like Billie Kompa used old Valentine motifs in new ways. There are still doves, roses, hearts, and forget-me-nots; but they are stylized and updated, c.1950. 15" cotton. $20-25.

This unique St. Patrick's Day handkerchief depicts a fine lady and gentleman playing tunes on an Irish harp, c.1950. 11.5" cotton. $15-20.

Everyone's Irish on St. Patrick's Day! The "wearin' o' the green" can be achieved with one of these bright green and white (and a touch of leprechaun's gold!) hankies, c.1960. 12" and 13" cotton. $12-15.

Tammis Keefe recalls the summer days spent at the shore. This beach scene uses outlandish colors like pea green, orange, khaki, black, and white with a framing rolled edge of orange, c.1950. 15" linen. $20-25.

Carl Tait takes a linear approach in this five color fall design. Using a grid pattern, white, gray, black, copper, and rust create a dramatic effect, c.1950. 15" linen. $20-25.

Fall is celebrated in the unusual colors of this Faith Austin design. The blue background compliments the turquoise and rust, which were so popular in the 1950s, c.1950. 15" linen. $20-25.

Brilliant colors against a black background cover this autumn design by Carl Tait. The central scarecrow dances amidst a shower of bright fall leaves, c.1950. 15" linen. $20-25.

A harvest theme of scarecrows in a cornfield is an original by Tammis Keefe. Purple and turquoise is an imaginative combination for this handkerchief, c.1950. 15" linen. $20-25.

"The Pilgrims" by Henry Charles is the perfect 1950s style remembrance of Thanksgiving Day. The red rolled edge provides a shock of color and surrounds the gray and pink, which were very popular at the time, c.1950. 15" linen. $25-30.

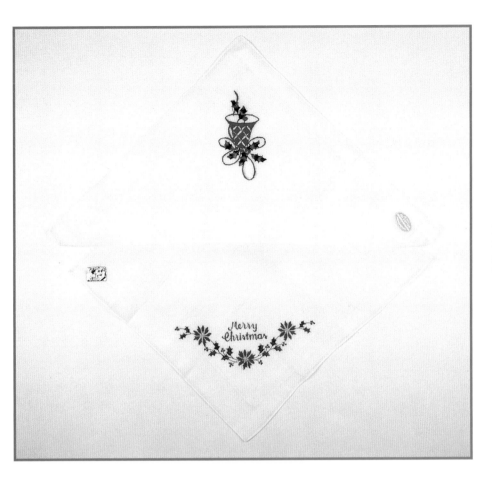

Fine cotton batiste with embroidered Christmas motifs. These handkerchiefs were often gifts for classmates or teachers, c.1950. 12" cotton batiste. $12-15.

Inexpensive printed cotton hankies made in Japan and the Philippines were produced toward the end of the 1960s. Original tags show that they sold for less than a dollar. Bright colors and bold designs still make these fun holiday gifts, c.1966. 13" cotton. $10-12.

The cut-out bows that decorate the Christmas wreaths on this handkerchief make it worth collecting, c.1950. 13.5" linen. $15-20.

Santa appears on these Christmas theme hankies. Not as common as floral designs, c.1960. 11" and 12" cotton. $20-25.

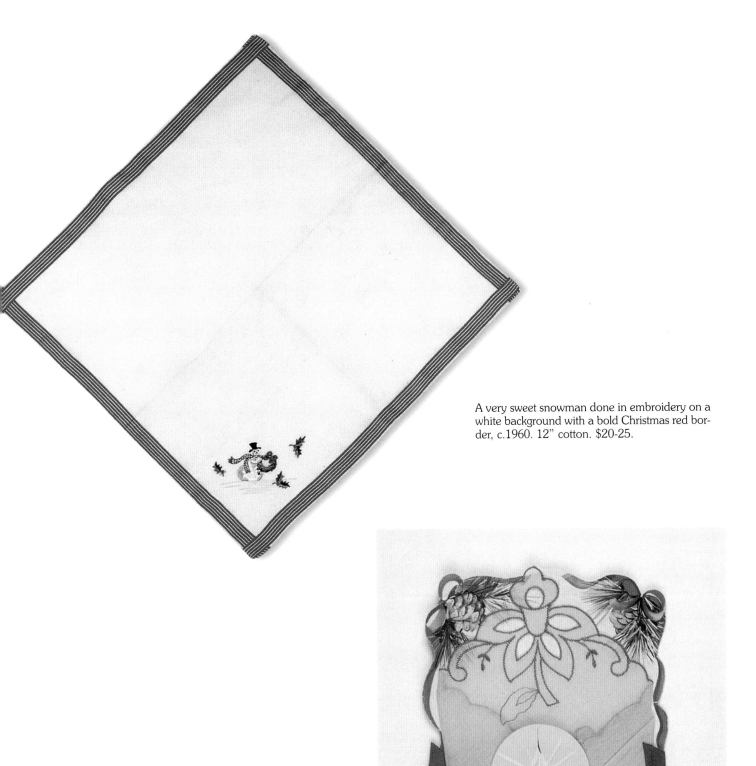

A very sweet snowman done in embroidery on a white background with a bold Christmas red border, c.1960. 12" cotton. $20-25.

This charming Christmas card holds a gift hankie—"To, Miss White/From, Betsy Read." Dated 12/25/54. 12" cotton. $20-25.

This vivid ruffled poinsettia is my personal Christmas favorite. The brilliant color and graphic quality make it very special, c.1950. 13" cotton. $25-30.

CHAPTER 4
ACROSS THE COUNTRY

The years following World War II brought new freedoms and prosperity to Americans. Wartime travel restrictions were dropped and all took to the road to celebrate! The handkerchiefs shown in this chapter are visual depictions of that new found freedom and sense of celebration. They are brightly colored souvenirs of the trips taken.

Famous cities and points of interest are recalled in these fabric mementos. Places like colonial Williamsburg, Atlantic City, and the Empire State Building are commemorated. America had become a nation of tourists, and trips across this great country became the theme of some highly collectible handkerchiefs. A series of handkerchiefs featuring state maps with state flowers and famous locations are of high interest among collectors. In fact, this chapter represents the largest area of concentration within our collection. Over thirty states are represented and several major cities including many popular tourist destinations.

What better place to begin, but at the top...of the Empire State Building, of course! The tallest building in the world at that time. Original label—Bloch-Freres, c.1950. 13.5" linen. *From the collection of Mary B. Ross.* $25-30.

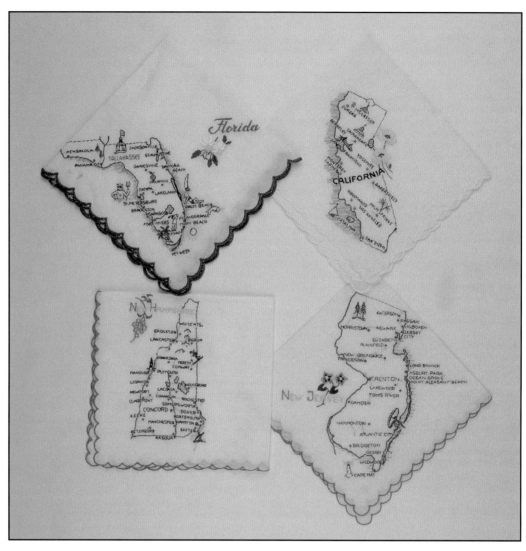

Embroidered maps of the states. These four have scalloped edges and feature the state flowers of Florida, California, New Jersey, and New Hampshire, c.1960. 11" cotton. $12-15.

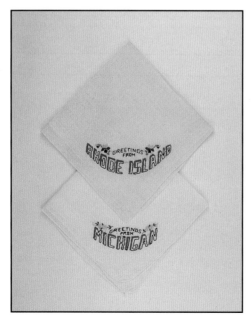

These seem to be men's handkerchiefs that bring greetings from Michigan and Rhode Island, c.1960. 12" cotton. $8-10.

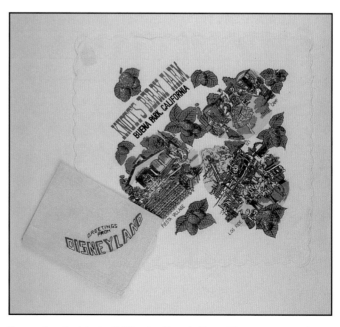

Souvenirs of a trip to California. Knott's Berry Farm is a cotton/rayon blend. Disneyland is a men's hankie, c.1960. 12" cotton. $8-10.

A lovely collection based on a trip West. Done in embroidery with lovely detail, c.1960. 11" cotton. $12-20.

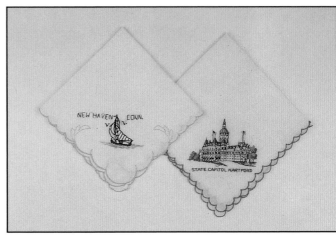

Above: Two hankies from the state of Connecticut. Embroidered, c.1960. 11" cotton. $12-20.

Left: Four embroidered examples from an East Coast trip. Two men's—Orchard Beach and Atlantic City. Two ladies'—The White House and Annapolis, c.1960. 12" cotton. $12-20.

Seeing the sights in New York City yielded these examples for the collection, c.1960. 11" cotton. $12-20.

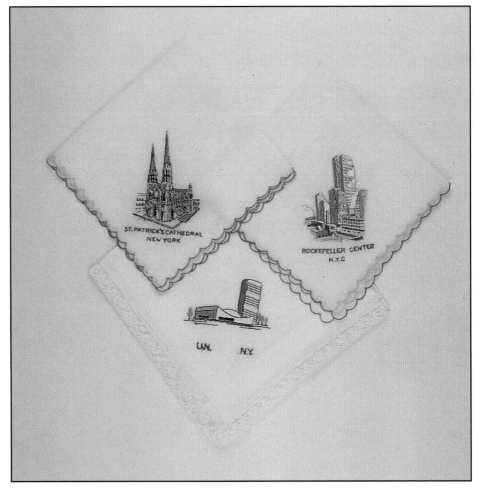

Inspired by a trip to Williamsburg, designer Emily Whaley uses brown and green on a yellow field to depict the colonial buildings of the town, c.1950. 14" linen. $20-25.

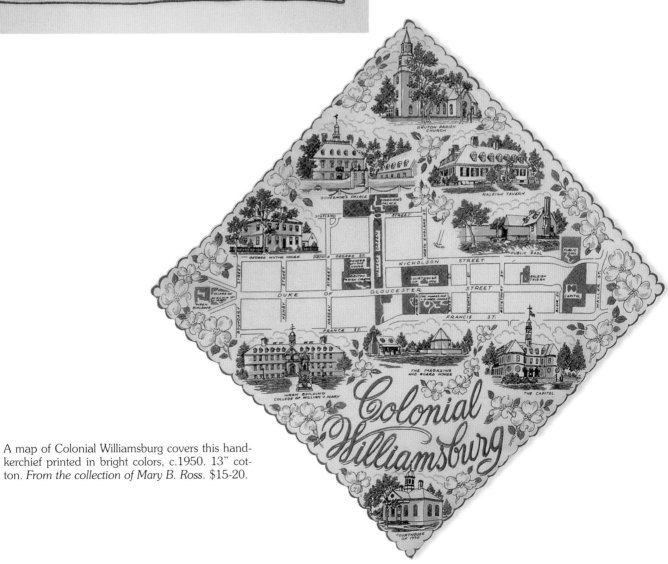

A map of Colonial Williamsburg covers this handkerchief printed in bright colors, c.1950. 13" cotton. *From the collection of Mary B. Ross.* $15-20.

Another Williamsburg inspired design by Emily Whaley, this time a pale blue background is decorated with darker blue and pink colonial motifs, c.1950. 14" linen. $20-25.

The highpoints of New England appear in this souvenir handkerchief. Attractive colors and designs make this a good item to collect, c.1950. 13" cotton. $15-20.

Pat Prichard produced this design based on the map of New England. With accents of lavender and chartreuse, c.1950. 14" linen. $20-25.

Two color versions of a New England harbor scene. Turquoise on white and brown on tan. Original label—Amanda Prints, c.1950. 13" linen. $20-25.

A day at the fair—in Orange County (California!), c.1950. 13" cotton. $12-15.

Denver—"The Mile High City"—
is celebrated in this appealing design based on points of interest.
Did you know Buffalo Bill was
buried there? Original label—
Burmel, c.1950. 13.5" cotton.
$18-20.

Designer Carl Tait does Dayton, Ohio. In poor condition, it appears because of the charming airplane in the center square, c.1950. 13.5" linen. $8-10.

A similar handkerchief by Carl Tait represents Harrisburg, Pennsylvannia. Original label—Herrmann Handkerchief, c.1950. 13.5" linen. *From the collection of Mary B. Ross.* $20-25.

An unusual circular hankie "Around Chicago" features the sights of the city, c.1950. 13" diameter cotton. *From the collection of Mary B. Ross.* $20-25.

Remember the song *Chicago, Chicago...That Wonderful Town?* Designer "don" does in this stylized map, c.1950. 13" cotton. $15-20.

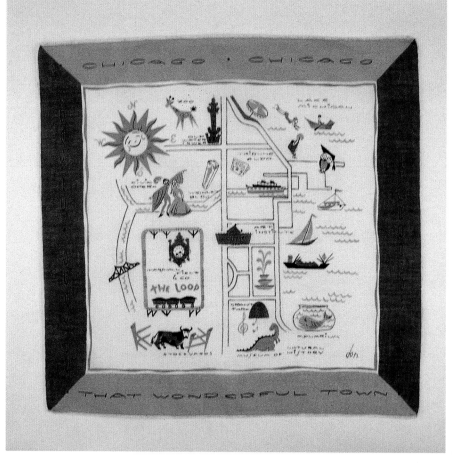

"The Windy City" is the theme again. Chicago map done in black and white on a blue field. Original label—Herrmann Handkerchief, c.1950. 13" cotton. *From the collection of Mary B. Ross.* $20-22.

A colorful tribute to Richmond, Virginia. It features a vintage plane at the Byrd Airport, c.1950. 13" cotton. $15-20.

Meet me in St. Louis, Missouri, c.1950. 13" cotton. $15-20.

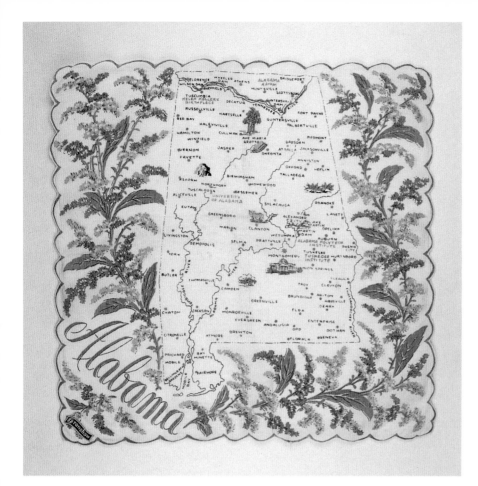

The part of the collection featuring printed maps of the states begins with Alabama. Done in blue, original label—Franshaw, c.1950. 13.5" cotton. $20-22.

Alaska map. Blue color theme, c.1950. 13.5" cotton. $18-22.

Arizona. Done in green on white ground, a very detailed map, c.1950. 12" cotton. *From the collection of Mary B. Ross.* $15-18.

Arizona. Very poor condition, c.1950. 13.5" cotton. $6-8.

Arkansas. Bill Clinton was probably born around the time this hankie was produced, c.1950. 13.5" cotton. *From the collection of Mary B. Ross.* $18-22.

California. Red poppies, c.1950. 13.5" cotton. *From the collection of
Mary B. Ross.* $18-22.

California, c.1950. 13" cotton. *From the collection of Mary B. Ross.* $18-22.

California. Red and blue versions of the same handkerchief, c.1950. 13.5" cotton. *From the collection of Mary B. Ross.* $18-22.

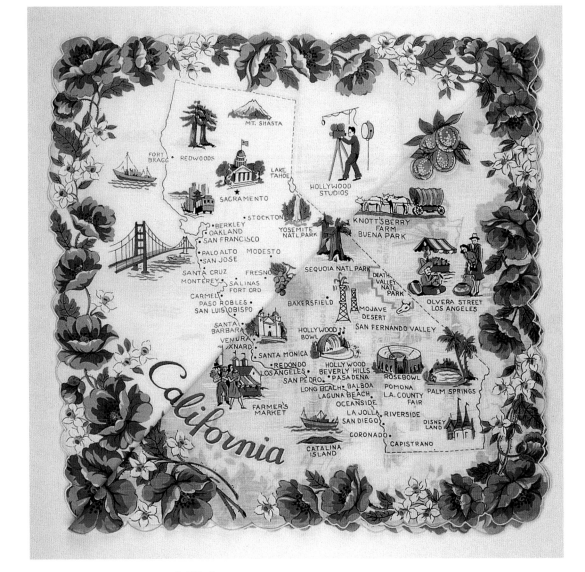

California. Pink background, c.1950. 13.5" cotton. $18-22.

California. Tan and brown with pink accents, c.1950. 14" cotton. $18-22.

California. Golden poppies, c.1950. 13.5" cotton. $18-22.

California. Oranges create the border and Mickey Mouse appears in the corner, c.1950. 14" cotton. $20-22.

California. Done in red and fuschia, two different versions of the same design, c.1950. 13.5" cotton. $18-22.

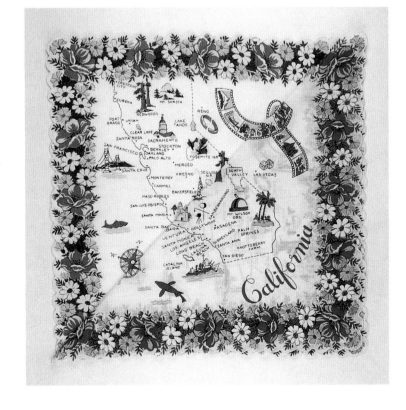

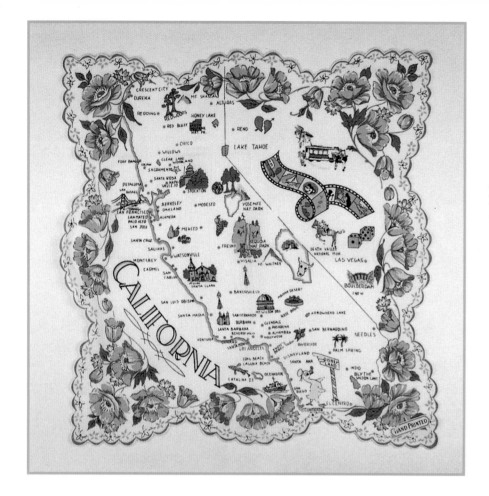

California. Blue and pink color scheme. Original label—"Hand Printed," c.1950. 13.5" cotton. $20-22.

California. A visitor marks their location with a "We're here" message to friends at home, c.1950. 13" cotton. $18-20.

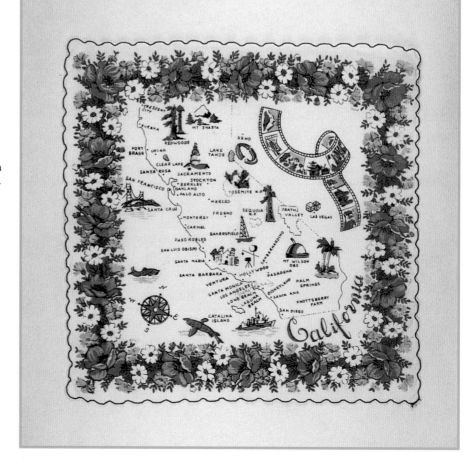

Colorado. Done in blue, c.1950. 13" cotton. $18-20.

Colorado. More details appear than in the first example, c.1950. 13" cotton. *From the collection of Mary B. Ross.* $18-22.

Florida. The state is printed in pink with a red border, c.1950. 13" cotton. *From the collection of Mary B. Ross.* $18-22.

Florida. More masculine colors—maroon, yellow, and orange—straight edge, c.1950. 13" cotton. *From the collection of Mary B. Ross.* $18-22.

Florida. The map in turquoise. Straight edge, c.1950. 13" cotton. *From the collection of Mary B. Ross.* $18-20.

Florida. Beautiful red hibiscus on a bold blue background, c.1950.
14" cotton. *From the collection of Mary B. Ross.* $18-22.

Florida. Miami is the focus, c.1950. 13" cotton. *From the collection of Mary B. Ross.* $18-22.

Florida. Original label—Made in Japan, c.1950. 12.5" cotton. $18-22.

Florida. A bit faded, c.1950. 13" cotton. $15-18.

Georgia. The same handkerchief as left, but faded from use, c.1950. 12" cotton. $15-18.

Georgia. Original label—Made in Japan, c.1950. 12" cotton. $20-22.

Illinois. Bunches of purple violets. Original label—Franshaw, c.1950. 13.5" cotton. $20-22.

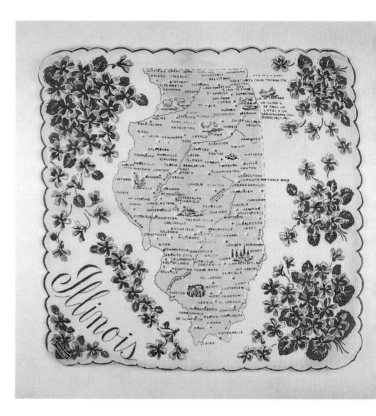

Hawaii. This handkerchief has a lovely hand painted look. Each corner contains a Hawaiian scene surrounded by swaying palms, c.1950. 12"cotton. $20-25.

Indiana. Some fading and few stains, c.1950. 13" cotton. $15-18.

Kentucky and Tennessee. These neighboring states share a hankie done in gold, c.1950. 13" cotton. $20-22

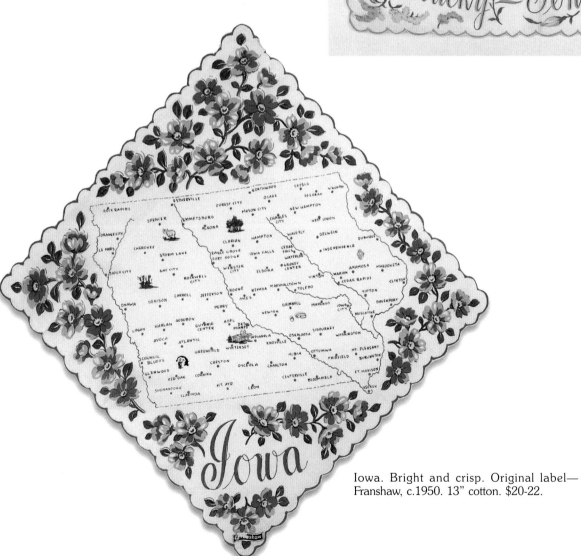

Iowa. Bright and crisp. Original label— Franshaw, c.1950. 13" cotton. $20-22.

Louisiana. Points of interest on a beige ground, c.1950. 13" cotton. *From the collection of Mary B. Ross.* $18-22.

Louisiana. Blue field. Original label—Franshaw, c.1950. 13" cotton. $20-22.

Massachusetts. Original label—Franshaw,
c.1950. 13" cotton. $20-22.

Massachusetts. "The Bay State," c.1950. 13" cotton.
From the collection of Mary B. Ross. $18-22.

Minnesota. Original label—Franshaw, c.1950. 13"
cotton. $20-22.

Mississippi. Original label—Franshaw, c.1950. 13" cotton. $20-22.

New Jersey. This may be faded, but it's a hard state to find, c.1950. 13.5" cotton. $20-22.

New York. "The Empire State," c.1950. 13"
cotton. *From the collection of Mary B. Ross.*
$18-22.

New York. Roses bloom on this faded
example, c.1950. 13" cotton. $15-18.

North Carolina. Pink blossoms, c.1950. 13"
cotton. $18-22.

Oklahoma. A bit of seam binding is attached to the edges of this hankie; it was going to be part of a handkerchief quilt, c.1950. 13" cotton. $18-20.

Pennsylvania. A commemorative of the discovery of oil in Titusville, Pennsylvania, c.1959. 13" cotton. *From the collection of Mary B. Ross.* $20-25.

Pennsylvania. Pink mountain laurel surrounds a simple map, c.1950. 13" cotton. $18-22.

Pennsylvania. A more detailed map, c.1950. 13" cotton. $18-22

South Dakota. Native American themes. Original label—All Cotton, Made in Japan, c.1950. 12" cotton. $20-25.

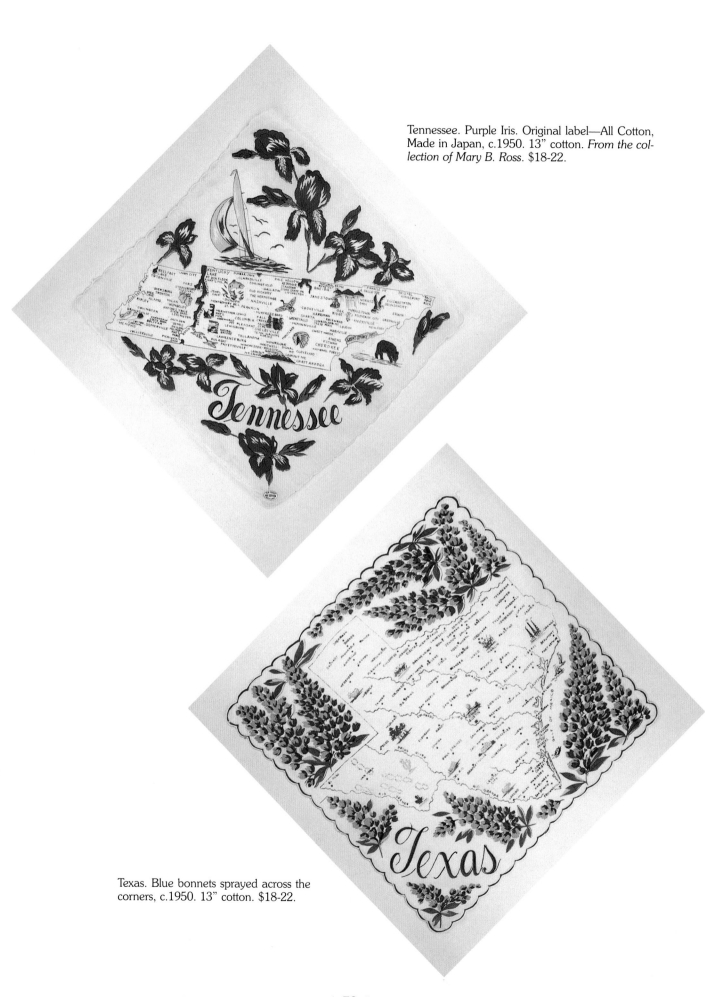

Tennessee. Purple Iris. Original label—All Cotton, Made in Japan, c.1950. 13" cotton. *From the collection of Mary B. Ross.* $18-22.

Texas. Blue bonnets sprayed across the corners, c.1950. 13" cotton. $18-22.

Vermont. A small state and hard to find, c.1950. 13" cotton. *From the collection of Mary B. Ross.* $18-22.

Virginia. Dogwood branches across the map. Original label—Franshaw, c.1950. 13" cotton. $20-22

Washington. "The Evergreen State," c.1950. 13" cotton. $18-22

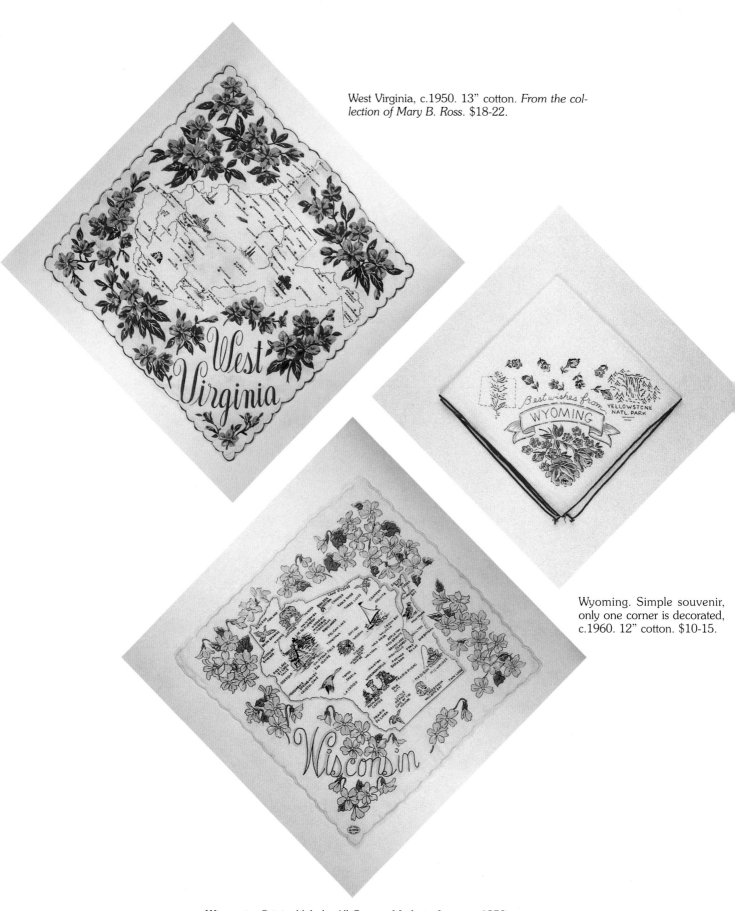

West Virginia, c.1950. 13" cotton. *From the collection of Mary B. Ross.* $18-22.

Wyoming. Simple souvenir, only one corner is decorated, c.1960. 12" cotton. $10-15.

Wisconsin. Original label—All Cotton, Made in Japan, c.1950. 12.5" cotton. *From the collection of Mary B. Ross.* $20-22.

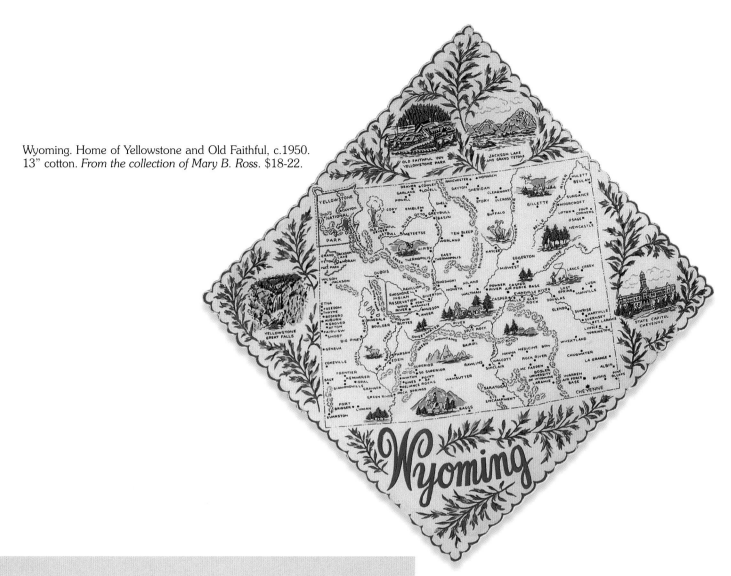

Wyoming. Home of Yellowstone and Old Faithful, c.1950.
13" cotton. *From the collection of Mary B. Ross.* $18-22.

Long Island. Home again. *Designed by Tammis Keefe.*
"The Yacht Harbors of Long Island," c.1950. 13.5"
linen. *From the collection of Mary B. Ross.* $20-25.

CHAPTER 5
AROUND THE WORLD

Many Americans began to travel as a hobby and cruises to exotic ports fulfilled the desire for new adventures. The love affair many Americans had with travel spread across the world. Trips to Europe and adventures in faraway places like New Zealand became possible with the increased availability of air travel. The handkerchief became the perfect souvenir—inexpensive, useful, decorative, fun, and easy to pack in a suitcase! They were available at every tourist shop, a customary stop on a traveler's itinerary. Collections were begun easily and took little room to store. Our collection reflects just how widely American's were traveling by the late 1950s with examples from every continent but Africa!

"Bon Voyage" by Tammis Keefe encourages travel by any means or mode, c.1950. 14" linen. $20-25.

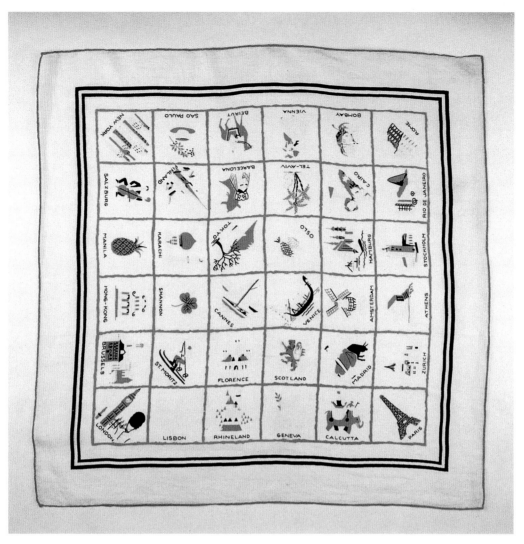

A pink, black, gray, and white color scheme enhances this travel theme handkerchief. Original label—Hand Rolled in the Philippines, c.1950. 14" linen. $20-25.

"Piccadilly Circus" in London, c.1950. 11" cotton. $10-15.

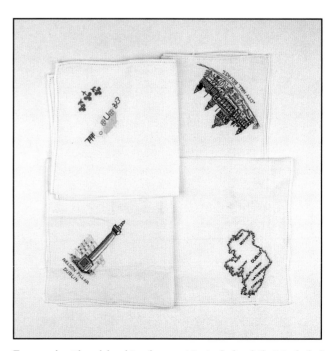

Four embroidered hankies from a trip to Ireland that included stops in Belfast and Dublin, c.1960. 10" Irish linen. $10-15.

"The Lakes of Killarney," c.1960. 12" Irish linen. $10-15.

"The German Wine Region." Great colors and detail in the lively map. Original label—Stoffel's, c.1960. 11" cotton. $12-15.

"How the Norwegians say it." Primary colors with a touch of green, c.1960. 11" cotton. $12-15.

"The Principality of Liechtenstein." A castle in the mountains. Original label—Kreier (signed), c.1960. 13" cotton. $12-15.

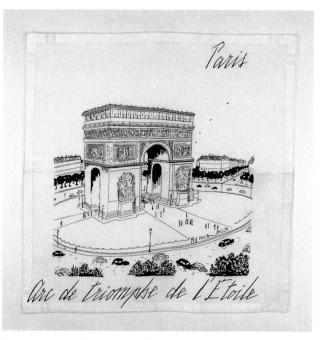

A pair in primary colors represent regions in Switzerland famous for its fine embroidery. Original label—Switzerland, All Cotton, c.1960. 11" cotton. $12-15.

France. Arc de Triomphe de l'Etoile" in black on white, c.1960. 11.5" cotton. *From the collection of Mary B. Ross.* $12-15.

France. "Arc de Triomphe" and other sights in Paris, c.1950. 15" linen. $15-18.

"How to Speak French"—a handy guide to important phrases. Done in a humorous hankie, c.1960. 13" linen. $18-20.

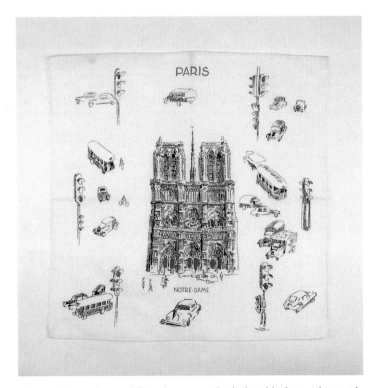

Paris. "Notre Dame." The famous cathedral in black-on-white with accents of red and green, c.1960. 12" cotton. *From the collection of Mary B. Ross.* $12-15.

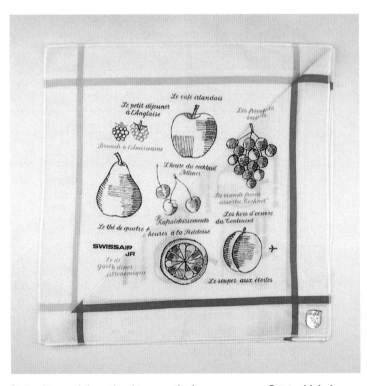

Swiss Air used these hankies as gifts for passengers. Original labels—Stoffel's, c.1960. 12" cotton. $12-15.

A cruise to the Bahamas provided these four embroidered views. Original label—Sun Dew, Pure Irish Linen, c.1960. 11" linen. $12-15.

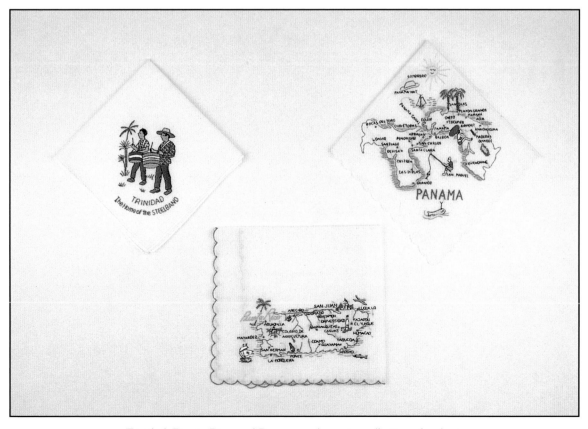

Trinidad, Puerto Rico, and Panama make a nice collection of embroidered souvenirs, c.1960. 11" cotton. $12-15.

South America. A gaucho hunts emu with a bollo, c.1950. 12" cotton. $12-15.

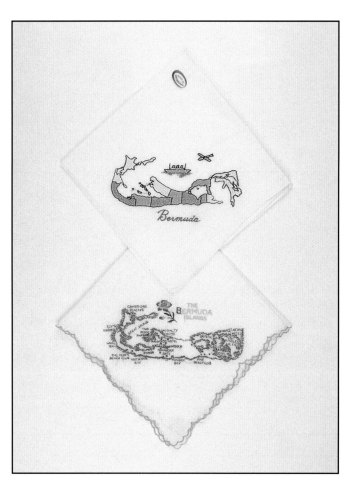

Bermuda. Two maps done in embroidery. Original label—Made in Switzerland, c.1960. 11" cotton. $12-15.

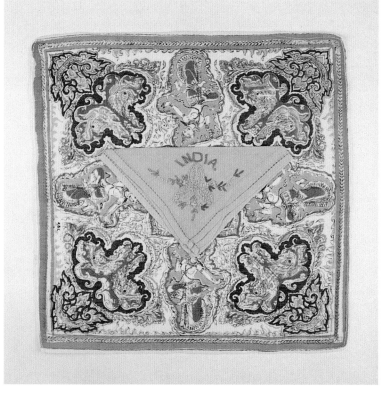

Two handkerchiefs that recall a journey to India. The pink embroidered is linen. The multi-colored design is cotton, c.1960. $12-15.

Montreal Canada. Pink chiffon, hand painted. White embroidered cotton, c.1960. $12-15.

What time is it? That all depends on where in the world you are! Moss green background with clocks from major cities around the globe, c.1950. 13.5" linen. $20-22.

A GROUP OF EIGHT HANDKERCHIEFS FORM A COLLECTION OF NEW ZEALAND SOUVENIRS. MAORI CULTURE, THE FAMOUS KIWI BIRD, AND TIKI STATUES ARE REPRESENTED. MAPS SHOW THE ISLANDS AND POINTS OF INTEREST, C.1960. 11" COTTON. $12-15.

CHAPTER 6
DESIGNERS

In the 1950s, a group of contemporary textile artists produced several series of handkerchiefs of outstanding quality. Collecting these is almost like collecting art. The small squares served as miniature canvases for the men and women who decorated them. They possess all the exuberance of the times, in their vivid hues and outrageous color combinations. The subjects used are imaginative and often whimsical.

This group is the newest addition to our collection. It has become my favorite due to the fine texture of the linen used and the unique themes that appear in the artists' designs. Done in a very "modern" style, these handkerchiefs represent all the best of the fabulous 1950s.

They all have a finely hand-rolled hem, often done in a boldly contrasting color. As a group, I feel they have the highest intrinsic value as they are the work of a pool of very talented artists. It seems likely that they could command the highest prices and will appreciate in value as time passes.

Pat Prichard. The perfect image of the 1950s—split level houses and television sets! The artist seems to be gently poking fun. Orange, turquoise, tan, black, and white, c.1950. 15" linen. $20-25.

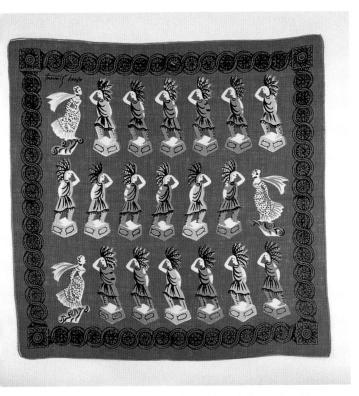

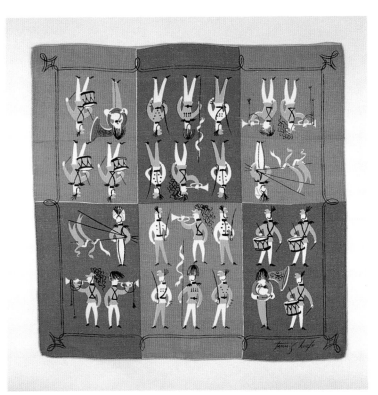

Tammis Keefe. Rows of wooden Indians face a lady made of the same stuff. Purple field, olive green trim, gray figures, and black and white accents, c.1950. 15" linen. $20-25.

Tammis Keefe. In alternating boxes of coral pink and bright blue, members of a marching band wear costumes of chartreuse, c.1950. 15" linen. $20-25.

Tammis Keefe. An African theme gets a silly twist with a smiling trophy moose in the center. Yellow, black, gray, orange, and white, c.1950. 15" linen. $20-25.

TAMMIS KEEFE. IN THIS SERIES OF THREE DESIGNS THE ARTIST HAS CHOSEN A POPULAR THEME, ANTIQUES! EACH HANDKERCHIEF HAS A CENTRAL MOTIF DONE IN BLACK AND WHITE, WHICH IS THEN SURROUNDED BY ROW UPON ROW OF OLD GLASSWARE, LAMPS, AND FURNITURE, C.1950. 15" LINEN. $20-25.

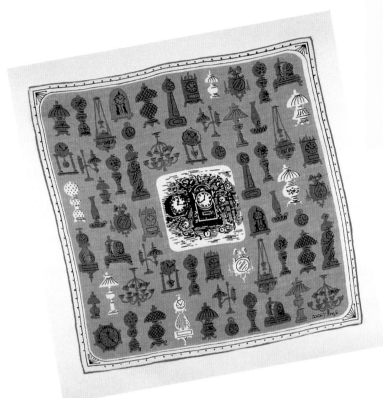

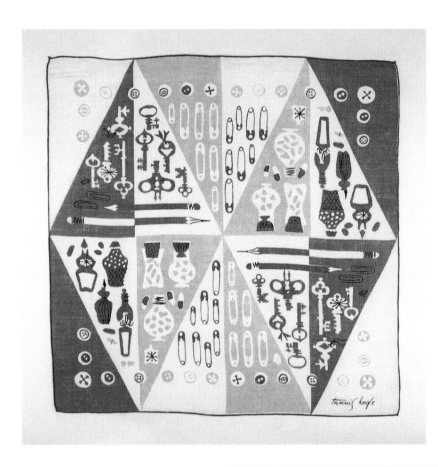

TAMMIS KEEFE. THE ART-
IST HAS DIVIDED HER
SPACE IN THE SAME PAT-
TERN IN BOTH HANDKER-
CHIEFS AND HAS FILLED
THOSE SPACES WITH ALL
THE LITTLE THINGS THAT
SEEM TO FILL MY KITCHEN
DRAWERS!. BLUE, TAN,
AND BROWN ARE THE
MAIN COLORS IN ONE—
THE OTHER COMBINES
PINK, GREEN, AND BLUE,
C.1950. 15" LINEN. $20-25.

Tammis Keefe. Intricate border detail creates an "Arabian Nights" feel with a hero on horseback as a center medallion. Hot pink, pink, olive green, black, and white, c.1950. 15" linen. $20-25.

Jeanne Miller. The colorful coins of many countries tumble across this handkerchief. An unusual blue background with turquoise, tan, olive green, black, and white, c.1950. 15" linen. $20-25.

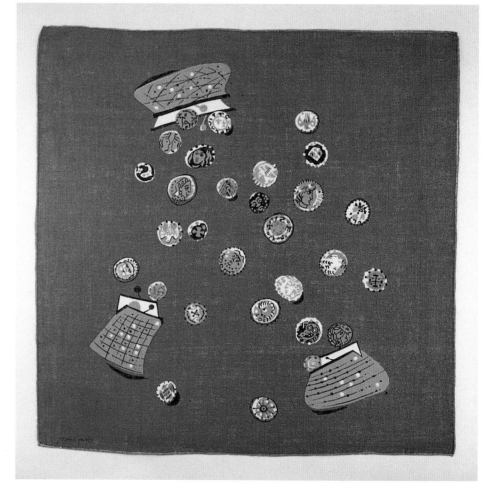

Jeanne Miller. Shelves of modern style home accessories done in brown, rust, peach, black, and white, c.1950. 15" linen. $20-25.

Pat Prichard. A lady's dressing table done in fantastic colors. Turquoise field, magenta, aqua, chartreuse, black, and white. Original labels: Hand Rolled—All Linen—Philippine Made. c.1950. 15" linen. $22-25.

Pat Prichard. Reading and writing are the themes here. Done in purple tones with accents of pink, black, and white. Original label—Franshaw, c.1950. 15" linen. $22-25.

Pat Prichard. An antique shop or an estate sale perhaps? Price tags appear on all the items on a deep purple field with pink, tan, black, and white, c.1950. 15" linen. $20-25.

Pat Prichard. Early American themes appear on a turquoise field. Olive green, magenta, black, and white accents, c.1950. 15" linen. $20-25.

Pat Prichard. Old steam engines chug across this design on white with red, blue, green, and black, c.1950. 15" linen. $20-25.

Pat Prichard. Clocks and cupids. Blue field, center of white, cherubs of lavender and violet with details done in black, c.1950. 15" linen. $20-25.

Pat Prichard. The old general store would have carried these items, but not in these crazy colors. Deep blue background with pink, yellow, brown, and white—black details, c.1950. 15" linen. $20-25.

Pat Prichard. Ice cream, umbrellas, and ladies' hats and gloves form the design on this olive green handkerchief with accents in gold and red, with touches of black and white, c.1950. 15" linen. $20-25.

Pat Prichard. An unusual copyrighted design done in lavender, turquoise, and olive green on a white handkerchief with black accents, c.1950. 15" linen. $22-25.

Frederique. Romantic forget-me-nots done in unique shades of tan, rust, and brown with yellow bows, c.1950. 15" linen. $20-25.

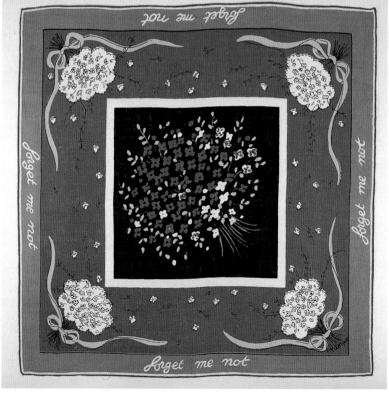

Frederique. Dramatic stripes in orange and black frame a woman dressed in blue and green, c.1950. 15" linen. $20-25.

Brigitta Ajnefors. Two charmingly detailed designs "Spring" and "Smörgasbord" have a very Scandinavian feel, c.1950. 15" linen. $20-25.

Carl Tait. Smiling suns and moons done in the classic 1950s colors of pink, black, and white, c.1950. 15" linen. $20-25.

Fragonard. "Fleurette," a potted plant with bright orange flowers, oversized, c.1950. 16" linen. $20-25.

Lori King. A design that plays on the artist's name includes two jacks, a queen, and a king. Turquoise, blue, olive, black, and white, c.1950. 15" linen. $20-25.

Monique. Artful feathers and leaves in a pastel palette—vivid pinks, blues, and greens on a soft pink field, c.1950. 15" linen. $20-25.

Ann McGann. A study of antique glassware in pink, red, blue, brown, and white with details in black, c.1950. 15" linen. $20-25.

Jean Sibo. A magical land with a hot air balloon drawn by birds in fantasy colors of green, blue, gray, and fuschia, c.1950. 15" linen. $20-25.

Faith Austin. A creative design with a Hawaiian theme. Grays, pink, and tan on a white handkerchief, c.1950. 15" linen. $20-25.

Kit Ann. This handkerchief is last because it represents a change. Look closely, the 1960s are coming with their psychedelic colors and paper tissues! Bright blue, pink, and yellow on a white handkerchief, c.1950. 15" linen. $20-25.

CHAPTER 7
RED, WHITE & BLUE

What is more American than the colors of our flag—red, white, and blue! These handkerchiefs are all based on that color scheme or have some other patriotic theme as a part of their design. Included in this chapter are the mementos soldiers and sailors sent home while away in the service. (A handkerchief was easy to post back to the states.) Pricing these touching souvenirs is very difficult, because they have such high sentimental value, so I have not included values for the few I have acquired.

Post-war patriotism made the use of the colors of the flag popular and positive. As a result, designers used the scheme effectively and proudly throughout the 1940s and 1950s in both clothing and accessories.

Tammis Keefe signed this handkerchief that features "The Declaration of Independence," c.1950. 15" linen. $20-25.

REPUBLICAN PARTY HISTORY

Intense opposition in nonslave states to further extension of the slavery system brought the Republican Party into being and it's first convention was held at Jackson, Michigan, July 6, 1854.

The first National Convention assembled in Philadelphia, Pa., June 17, 1856, and Abraham Lincoln was it's first successful standard bearer, being elected President of the United States in 1860 with Hannibal Hamlin as Vice President.

Fourteen Republican Presidents have served the United States since the "Grand Old Party" was first organized.

G.O.P. REPUBLICAN PARTY

REPUBLICAN PRESIDENTS

	BORN	BIRTHPLACE	ELECTED FROM	DIED	BURIAL PLACE
ABRAHAM LINCOLN	Feb. 12, 1809	Hardin Co. Ky.	Illinois	April 15, 1865	Springfield, Ill.
ANDREW JOHNSON	Dec. 29, 1808	Raleigh, N.C.	Tennessee	July 31, 1875	Greenville, Tenn.
ULYSSES S. GRANT	Apr. 27, 1822	Point Pleasant, Ohio	Illinois	July 23, 1885	New York, N.Y.
RUTHERFORD B. HAYES	Oct. 4, 1822	Delaware, Ohio	Ohio	Jan. 17, 1893	Fremont, Ohio
JAMES A. GARFIELD	Nov. 19, 1831	Orange, Ohio	Ohio	Sept. 19, 1881	Cleveland, Ohio
CHESTER A. ARTHUR	Oct. 5, 1830	Fairfield, Vt.	New York	Nov. 18, 1886	Albany, N.Y.
BENJAMIN HARRISON	Aug. 20, 1833	North Bend, Ohio	Indiana	Mar. 13, 1901	Indianapolis, Ind.
WILLIAM McKINLEY	Jan. 29, 1843	Niles, Ohio	Ohio	Sept. 14, 1901	Canton, Ohio
THEODORE ROOSEVELT	Oct. 27, 1858	New York, N.Y.	New York	Jan. 6, 1919	Oyster Bay, N.Y.
WILLIAM H. TAFT	Sept. 15, 1857	Cincinnati, Ohio	Ohio	Mar. 8, 1930	Arlington, Va.
WARREN G. HARDING	Nov. 2, 1865	Corsica, Ohio	Ohio	Aug. 2, 1923	Marion, Ohio
CALVIN COOLIDGE	July 4, 1872	Plymouth, Vt.	Mass.	Jan. 5, 1933	Plymouth, Vt.
HERBERT C. HOOVER	Aug. 10, 1874	W. Branch, Iowa	Calif.		
DWIGHT D. EISENHOWER	Oct. 14, 1890	Denison, Texas	New York		

The Republicans advertised their party on this hankie that lists Republican Presidents up to Eisenhower, c.1950. 13.5" cotton. $22-25.

This design features stars and a banner-like border, c.1950. 12" cotton. $10-12.

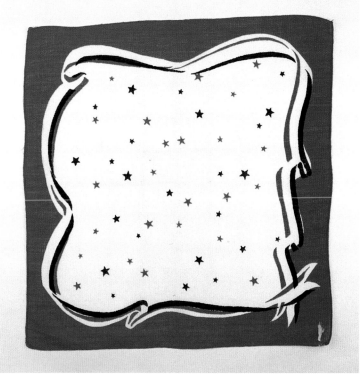

Military insignia and badges. I have seen this item
in shades of blue, c.1950. 13" cotton. $12-15.

Red and white. Blue and white, c.1950. 12" cotton. $8-10.

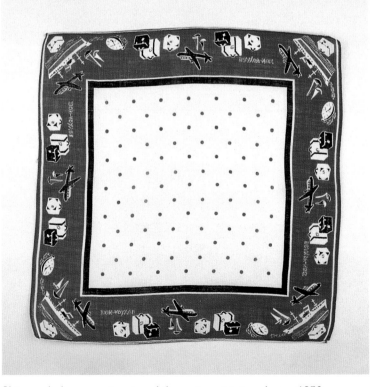

Ships and planes create a travel theme in patriotic colors, c.1950.
11" cotton. $12-15.

Stars and stripes, c.1950. 11" cotton. $10-12.

Red and white floral. Original label—Created by Kimball, c.1950. 12" cotton. $10-12.

"V for Victory" a wartime theme, c.1940. 12.5" cotton. $20-22.

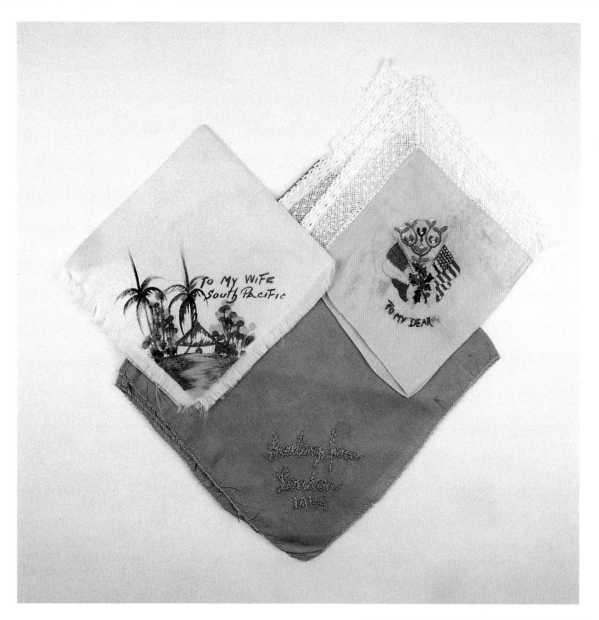

This group of handkerchiefs were sent home from servicemen. Touching reminders of long ago, they were often handmade by the men themselves, c.1940. 12" silk and satin. No price available.

The Army and the Navy are honored in these souvenir handkerchiefs, c.1950. 10" cotton and silk. $10-12.

The music swirls around this marching band in a bold and interesting design with a musical theme, c.1950. 13" cotton. $15-18.

Advertisers used the red, white, and blue often. Here a commemorative of the 100th consecutive broadcast of the *Woolworth Hour* on CBS Radio Network. Handkerchiefs seemed to lose favor around the same time radio began to be replaced by television, c.1957. 14" cotton. *From the collection of Mary B. Ross*. $25-30.

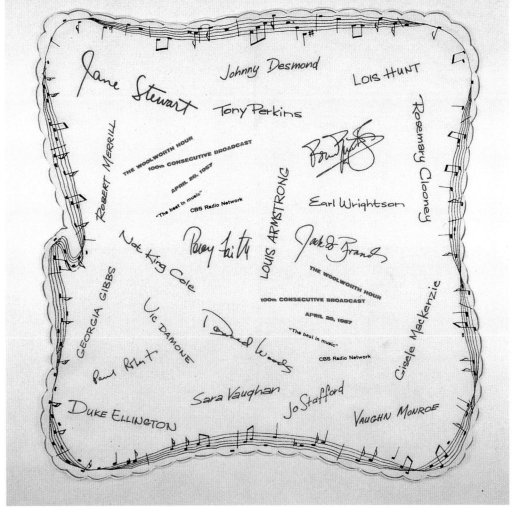

CHAPTER 8
FLORALS

The floral handkerchief is probably the most common type of collectible handkerchief. This may make them somewhat less valuable monetarily, but they remain a perennial favorite nonetheless. Their bright colors and lively prints endear them to all. These were our ordinary every-day hankies, nothing fancy, but still pretty.

Hankerchiefs of this sort are the easiest to find, so search for a favorite flower or save an unusual color. Use them and enjoy them. Tuck them in a denim shirt pocket or frame a group in a bedroom. Their charm is enduring and they never fail to bring a smile to my face when I look at them.

Boldly colored tulips in red and pink, cut out from a beige background, c.1950. 14" linen. $12-15.

Salmon colored hibiscus cut out from a turquoise field, c.1950. 14" linen. $12-15.

Full-blown roses in red, yellow, and even blue, gray border, c.1950. 12" cotton. $8-10.

Still glued to its original box paper. Purple tulips with a pink border, c.1950. 12.5" cotton. $10-12.

PROBABLY SOLD AS A SET, THESE THREE HANKIES HAVE WIDE SOLID COLOR BORDERS WITH LARGE CENTRAL SPRAYS OF VIVIDLY COLORED FLOWERS, C.1950. 12.5" COTTON. $10-12.

ALSO OBVIOUSLY FROM A SET, THE DEEP BLUE BACK-GROUND OFFERS A SHARP CONTRAST TO THE RED, WHITE, YELLOW, AND GREEN STRIPES AND TINY FLOWERS, C.1950. 13" COTTON. $10-12.

Lovely golden lilies with turquoise leaves and a brown, yellow, and gold border, c.1950. 12" cotton. $8-10.

Dayton's five and dime store charged 25¢ for this mint green bordered hankie. Original price tag, c.1950. 12" cotton. $8-10.

Two versions of the same hankie. Both color ways show an Oriental influence. Original price tag—Dayton's 25¢, c.1950. 12" cotton. $8-10.

All-over bouquets on a black background, white ribbon around a green border. Original price tag—Dayton's 25¢, c.1950. 12" cotton. $8-10.

A bright blue field with orange and black lilies. Original price tag—Marshal Field's 25¢, c.1950. 13" cotton. $8-10.

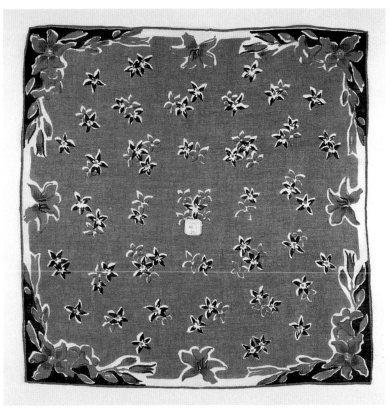

A central ground of a deep azure blue encircled by sprays of gold and green flowers with a deep brown border. Original price tag—Dayton's 25¢, c.1950. 12" cotton. $8-10.

An unusual deep maroon center wreathed with flowers of dark pink and blue. Polka dots and stripes in white complete the design. A soft scallop edge, c.1950. 13" cotton. $10-12.

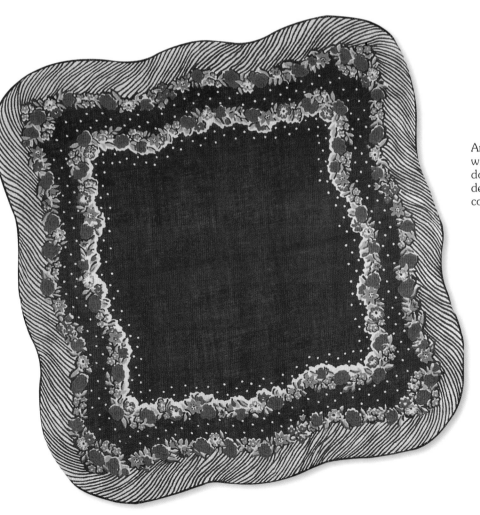

Brilliant red and pink tulips flow across a field of deep purple. Contrasting rolled edge of red, c.1950. 13" cotton. $10-12.

Sheer stripes with vivid black, purple, and fuschia daisies, c.1950. 12" cotton organza. $10-12.

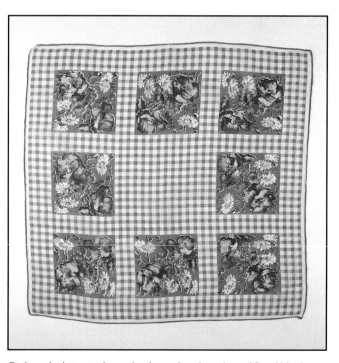

Pink and white gingham checks enclose bunches of floral blocks, c.1950. 13" cotton. $8-10.

A bouquet of flowers serves as a thank you and carries congratulations and best wishes from the sender, c.1960. 13" cotton. $8-10.

Pansies of red, pink, and yellow are bordered in a bold blue, c.1950. 12" cotton. $8-10.

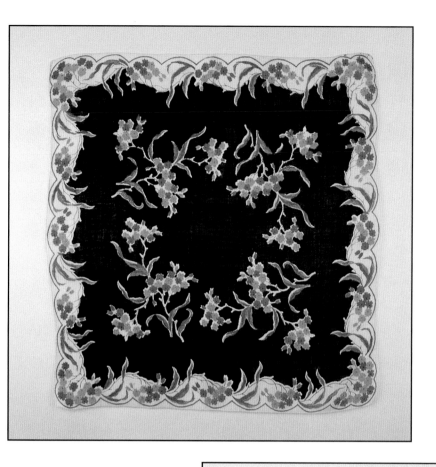

Black backgrounds were popular for the sharp contrasts that they offered. Used here with orange and brown blossoms, c.1950. 12" cotton. $10-12.

Dramatic red day lilies cut out from a black field, c.1950. 13" cotton. $12-15.

Sunny yellow lights up orange day lilies in this lovely and realistic design, c.1950. 14" cotton. $12-15.

Made in Switzerland with original label. Embroidery is used to produce bunches of flowers on white, c.1960. 12.5" cotton. $12-15.

A variety of Swiss embroidered handkerchiefs all with floral motifs found in one corner, c.1960. 12" cotton. $8-10.

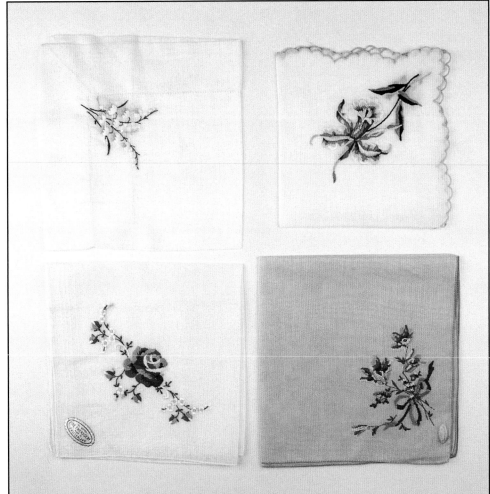

CHAPTER 9
RUFFLES & ROUNDS

There is nothing quite so pretty as the ruffled edges of this group of handkerchiefs. Cut in intricate scallops and hemmed by machine, they are highly feminine. The curves and points that edge these designs add an extra touch to the already lovely floral patterns that cover them. Some discerning collectors search only for these ultra-fancy hankies, picking through piles to find that special one!

Perhaps the most unusual type of handkerchief is the *round*. Produced briefly in the late 1940s and into the 1950s, the shape wasn't very popular at that time. It is, however, popular with present-day collectors who love to find these uniquely shaped hankies. It is interesting to see the clever designs that textile artists created on them.

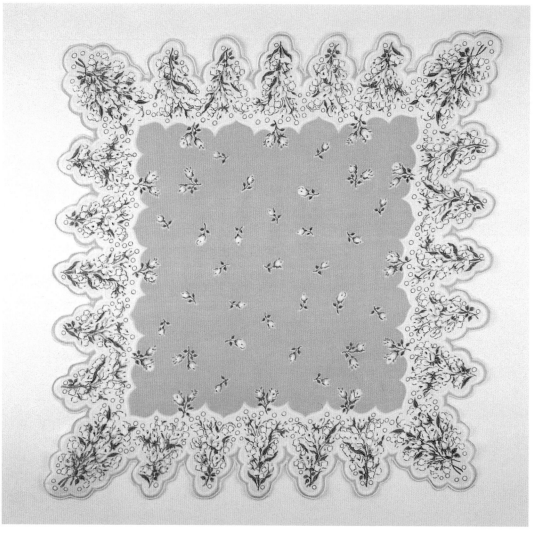

The lacey cut edges of this handkerchief are covered in tiny white rose buds and lily-of-the-valley, c.1950. 13.5" cotton. $15-18.

Printed lace bands edge this spring green handkerchief with lily-of-the-valley covering the center ground, c.1950. 13.5" cotton. $15-18.

The crisp white scallop edge contrasts well with the brown background of this daisy and petunia design. Original label—Brimel Original, c.1950. 13" cotton. $15-20.

Another lily-of-the-valley print with a blue background framed by bands of lighter blue. Original labels tell us it was made in the Philippines, c.1950. 13.5" cotton. $15-20.

Pink orchids on deeply cut gray leaves. A fabric flaw and some wear, c.1950. 12.5" cotton. $8-10.

Dogwood blossoms on an apple green background, c.1950. 13" cotton. $12-15.

Red carnations on a lacey border of black and white. A bit of edge wear, c.1950. 13" cotton. $8-10.

Golden cut out leaves support a colorful bouquet of pink poppies, white tulips, and blue forget-me-nots, c.1950. 13" cotton. $12-15.

Roses done in metallic gold give a very elegant touch to this handkerchief with a gray center field, c.1950. 13" cotton. $12-15.

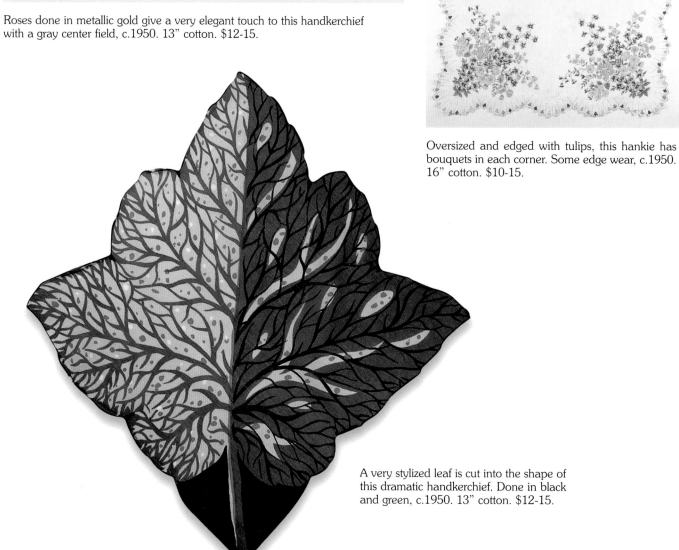

Oversized and edged with tulips, this hankie has bouquets in each corner. Some edge wear, c.1950. 16" cotton. $10-15.

A very stylized leaf is cut into the shape of this dramatic handkerchief. Done in black and green, c.1950. 13" cotton. $12-15.

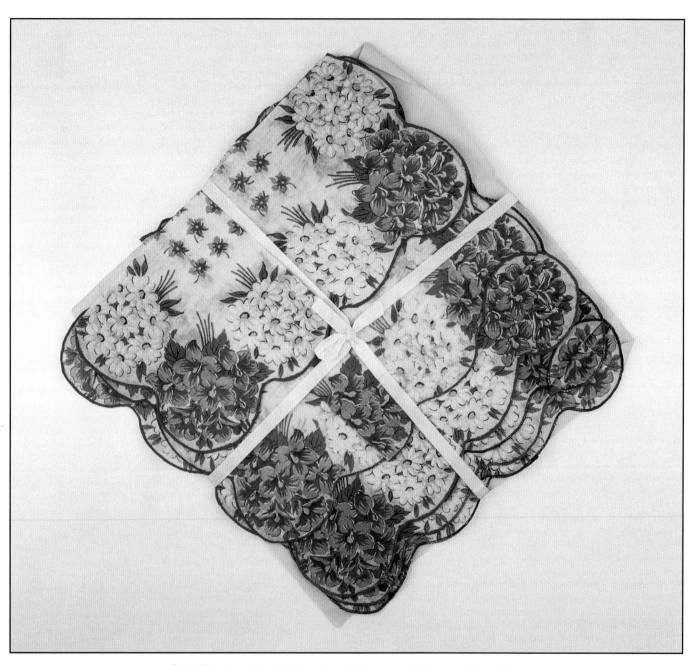

Set of five boxed ruffled handkerchiefs; two red, blue, purple, and turquoise. Sweet flowers accented by daisies, c.1950. 13" cotton. $55-60.

Bunches of rose buds bloom all over this lively red ruffled hankie, c.1950. 12" cotton. $10-12.

A sophisticated deep blue crimped edge with lighter blue embroidered dots on a pale blue fabric, c.1950. 12" cotton batiste. $12-15.

Daisies done in white with lavender accents border a white center ground. Original label—Made in the Philippines," c.1950. 13" cotton. $15-18.

Vivid contrast between a black center field and white daisy filled border, c.1950. 13" cotton. $12-15.

The first round handkerchief found for the collection is bright green with red roses encircling the border, c.1950. 13" cotton. $15-20.

Circles on circles—red, gray, blue, and black dots on a round handkerchief, c.1950. 12.5" cotton. $15-18.

Two large pink roses with turquoise daisies are surrounded by a border of pink daisies with a turquoise scalloped edge, c.1950. 11.5" cotton. $15-18.

A wreath of blue roses encircles a field of blue rose buds. Original label— Kreier, c.1950. 13" cotton. $15-18.

Sprays of red roses on a field of white. Original label—Stoeffel's Made in Switzerland, c.1950. 12.5" cotton. $15-18.

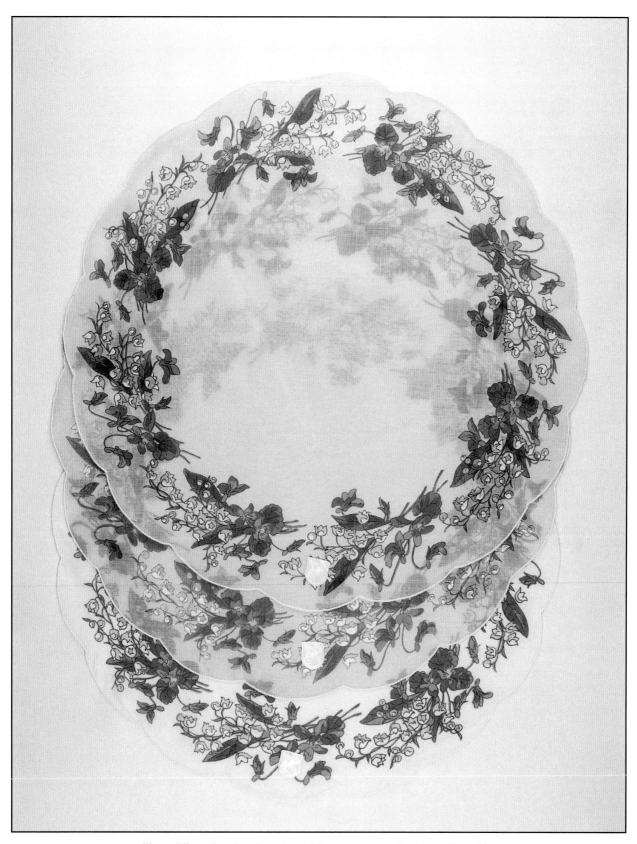

Three differently colored versions of the same design. Bunches of lily-of-the-valley and violets decorate the border. Original label—Stoffel's, Made in Switzerland, c.1950. 12.5" cotton. $15-18 each.

Circles of daisies, one of pink and one of blue, in these two examples. Original label—Krier, c.1950. 13" cotton. $15-18 each.

Two round florals show how charming wreaths of flowers can be on a simple circle. Original label—Stoffel's, Made in Switzerland, c.1950. 12" cotton. $15-18 each.

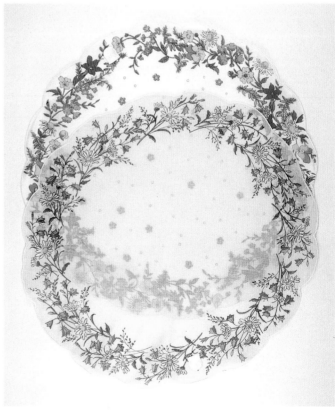

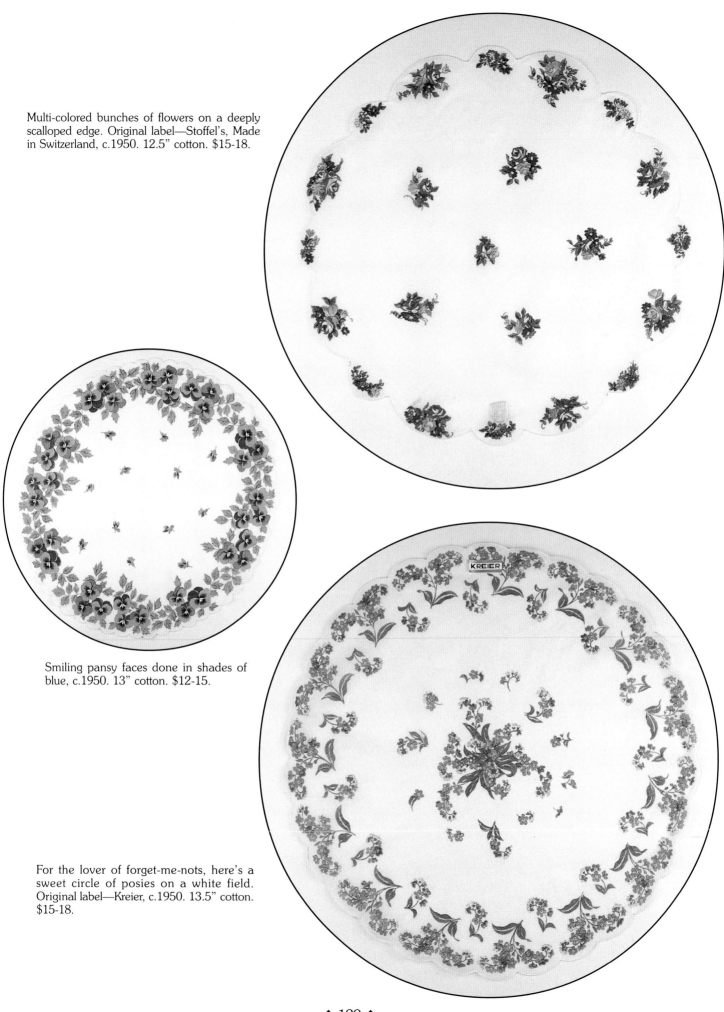

Multi-colored bunches of flowers on a deeply scalloped edge. Original label—Stoffel's, Made in Switzerland, c.1950. 12.5" cotton. $15-18.

Smiling pansy faces done in shades of blue, c.1950. 13" cotton. $12-15.

For the lover of forget-me-nots, here's a sweet circle of posies on a white field. Original label—Kreier, c.1950. 13.5" cotton. $15-18.

"Around Hartford"—Souvenir of the Connecticut city with the state flower as a center medallion, c.1950. 13.5" cotton. $18-20.

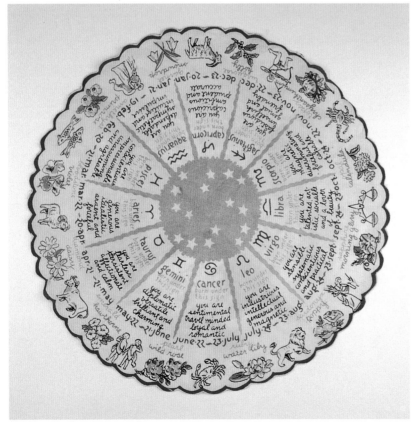

A very interesting design with the twelve signs of the zodiac, a few stains, c.1950. 13" diameter cotton. $18-20.

A very attractive roulette wheel done on a round hankie. What a great design for this type of handkerchief, c.1950. 13.5" cotton. $20-22.

Lovely gray and white leaves swirl around butterflies of pink, white, and chartreuse of a light gray field, c.1950. 13" cotton. $15-18.

Very large round wreath with daisies and sashed with bows of pink. Desirable because of its unusual size, c.1950. 17" cotton. $20-25.

An entire rose done on a round hankie. Vivid shades of pink make this a very special addition to a collection, c.1950. 11" cotton. $20-25.

This spectacular purple flower bursts into full bloom on this round design, c.1950. 11" cotton. $20-25.

CHAPTER 10
SPORTS & ANIMALS

Design themes for handkerchiefs range far and wide. Some of the most unique and charming artwork is found on hankies that feature sports and animals design motifs. Designers seemed to enjoy the things we also enjoy—our pets and recreational activities.

Often designed to appeal to a younger crowd, or at least the young at heart, these handkerchiefs are often humorous and frequently down right silly. The sometimes-garish colors, popular during the 1950s, added a bit of fantasy to everything they touched, making vintage handkerchiefs with animal and sports themed designs very appealing to current collectors.

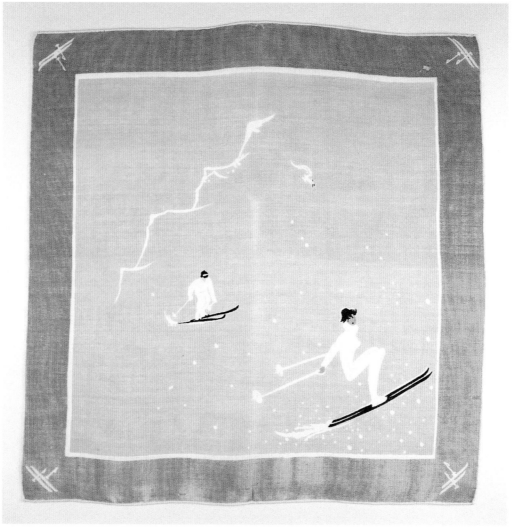

A gray frame surrounds a pink mountain with an attractive
"snow bunny" skiing by, c.1950. 15" linen. $20-25.

The same sporty young woman appears with a group of friends on bicycles. The vivid red border contrasts with a gray background, c.1950. 15" linen. $20-25.

Horses and riders in brown, tan, and white on a blue field are enclosed by a brown fence in this hunt scene, c.1955. 13.5" linen. $20-25.

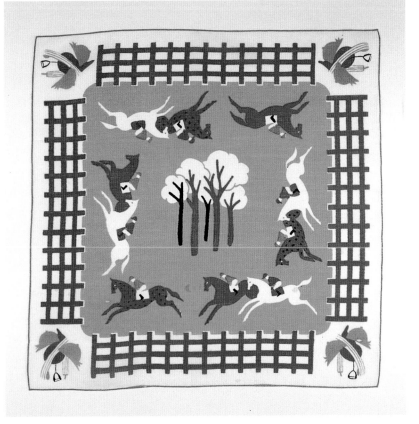

Probably designed to encourage young men to become athletes, this five color design shows five sports: boxing, hiking, archery, fencing, and running. The red border provides a bright contrast to the white background color, c.1950. 13" cotton. $18-25.

What a catch! This lucky fisherman has a mermaid on his line. Good luck in any language, c.1960. 11" cotton. $18-25.

The fishing creel, flies, and rod are embroidered on this white handkerchief, c.1950. 13" cotton. $18-20.

This humorous hankie by Pat Prichard implies that ducks really don't like rain. The umbrellas and ducks are pink, black, orange, and white on a light blue ground, c.1950. 15" linen. $20-25.

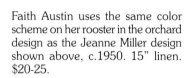

Jeanne Miller uses fantasy colors on this carousel horse. A palette of greens and blues enlivens this design, c.1950. 15" linen. $20-25.

Faith Austin uses the same color scheme on her rooster in the orchard design as the Jeanne Miller design shown above, c.1950. 15" linen. $20-25.

A true pampered pet—with a whimsical message for the collector. "So?" designed by Hazel Ware. Pink, yellow, gray, white, and black, c.1955. 15" linen. $20-25.

Another of Hazel Ware's "fat cats" (as seen above), this time we see a gentleman with jewels for his pussycat, c.1955. 15" linen. $20-25.

The bold black background of this Tammis Keefe design provides contrast to the white tree and animals in green, blue, orange, and brown, c.1950. 15" linen. $20-25.

Embroidered flying pheasants appear in each corner of this white handkerchief. The appliqued and embroidered turtle looms large in his corner, c.1960. 12" cotton. $10-15.

A sophisticated tiger print border surrounds a charming tiger with an umbrella. This whimsical design is by Tammis Keefe, c.1950. 15" linen. $20-15.

Pat Prichard does the "pampered pooch" in this design based on a day in the life of an elegant French poodle. Gray background with black, pink, and gold dogs, c.1950. 15" linen. $20-25.

"Don't look a gift horse in the mouth" is the theme of this humorous handkerchief done in pink, red, chartreuse, and gold. A small stain and remains of the original paper label, c.1950. 15" linen. $18-25.

"A day at the races"—done by Tammis Keefe. Three boldly colored bars of jockeys and horses. Orange, green, chartreuse, and black, c.1950. 15" linen. $20-25.

A classic rendition of a horse and rider going over a jump. Original fabric tag—a. skandia print Made in 1966-67. 12" cotton. $15-20.

Another classic horse scene. A blue background with mare and foal done in black and white. Original fabric tag—a. skandia print (1964). 12" cotton. $15-20.